HOW TO GET A FARM,

AND

WHERE TO FIND ONE.

SHOWING THAT

HOMESTEADS MAY BE HAD BY THOSE DESIROUS OF SECURING THEM:

WITH

THE PUBLIC LAW ON THE SUBJECT OF FREE HOMES, AND SUGGESTIONS FROM PRACTICAL FARMERS;

TOGETHER WITH

NUMEROUS SUCCESSFUL EXPERIENCES OF OTHERS, WHO, THOUGH BEGINNING WITH LITTLE OR NOTHING, HAVE BECOME THE OWNERS OF AMPLE FARMS.

BY THE AUTHOR OF

"TEN ACRES ENOUGH."

New York:
PUBLISHED BY JAMES MILLER,
(SUCCESSOR TO C. S. FRANCIS & CO.)
522 BROADWAY.
1864.

RENNIE, SHEA & LINDSAY,
Stereotypers and Electrotypers
81, 83 & 85 Centre-street,
New York.

Anderson & Ramsay,
Printers,
28 *Frankfort Street, N. Y.*

PREFACE.

THE rich man needs no such work as this. His ample purse will enable him to purchase land where-ever his fancy may lead, paying for other men's improvements, and lavishly expending his means on new ones. He has his idols in common with the poor man. The first thought of the former is to improve and embellish; that of the latter is simply to acquire.

The now wealthy man was at one time actuated by a similar impulse. Henceforth his ambition is to spend. As the poor are always with us, there is a constantly existing crowd whose aspirations are identical with those which he once entertained. Many of them are equally deserving with their successful predecessors. Many of them have no thought of achieving fortune by commerce, trade, or manufactures, or the national vice of office-seeking. Their attention is directed exclusively to agriculture, and the acquisition of land. They have either been brought up as farmers, or a passion has been born with them to become such, or disappointment elsewhere has turned their thoughts in the same direction.

In all these cases, they are aiming for a common goal—the securing of a farm. Multitudes succeed in their object, while other multitudes fail—some from ignorance, some from incurable incapacity, others from misdirection. The man who digs for gold at random will invariably become poor, while he to whom the precise spot has been pointed out wherein the precious deposit lies concealed, will, with a fraction of the same industry, become rich. To be successful in any thing, effort must be directed by intelligence. Fortunes may be stumbled on occasionally, but stumbling will be found to be a very precarious dependence.

So far as misdirection may be a cause of failure, it can to some extent be avoided. My object is to show how such result may be prevented, by suggesting practical methods for insuring success—some original, some derived from the ripe experience of others. I write with no reference to mere land speculation, such as induces men to purchase to-day for the sole object of selling at a higher price to-morrow, the new buyer selling a week later to a still newer one, while neither has, in the interval, expended a dollar in improvements. I treat almost exclusively of gradual increase of value, and only incidentally of sudden enhancement. Incidents of the latter do occur without the owner's having ever contemplated them. While not to be disregarded as incidentals, they are not adopted as primaries.

My effort has been to group together in the following pages some of the many remarkable openings for agricultural enterprise which exist in our country. Wherever we turn they are to be found. The great West has long abounded with them, and the South will soon be equally prolific. The Middle States, New Jersey, Delaware, and Maryland, contain thousands of these openings, where cheap lands within reach of cash markets have long been waiting for purchasers. But they have remained comparatively unknown to the agricultural public. The owners have not prized them as they deserved to be, and the speculators have overlooked them. The great West has carried off the honors as well as the population.

It is believed that an acceptable service will be rendered to inquirers, by bringing together, in a single compact view, a description of these several classes of openings. By thus having them in a hand-book, they can be readily and conveniently examined. Each inquirer can read and determine for himself. The variety may be pronounced confusing. No other country offers a tithe of the inducements that are held out to all classes in this. Wherever a man may incline to settle, there some eligible opening will be found to exist, no matter whether he contemplates engaging in agriculture or not. In endeavoring to show all how to get a farm, it was important to inform them where it might be had. On both points they will here find abundant information;—the action must be taken by themselves.

An effort has been made to draw attention to the great but unappreciated value of the numerous tracts of swamp-lands which are to be found among the centres of population in all the older States. The subject might have been further elaborated by suggesting the application of organized capital to this enterprise on a large scale. It has been thus organized and applied in Europe; but our country is probably too young, and land too abundant, for an extensive undertaking of that character to be entertained.

Particular reference has been made to the vast quantities of cheap lands for sale in the three States of New Jersey, Delaware, and Maryland. The information touching these lands and their productions, has been derived, in some instances, from correspondents on the spot. In others, as in Delaware and New Jersey, my account is mainly from personal inspection. I could reach them conveniently, and had the fullest opportunity for making a very thorough examination. I conversed with many persons who had settled there from other States, saw their improvements, as well as their crops, and received candid replies to all inquiries as to how they liked their new locations, and how they were succeeding. The facts thus acquired are reported without suppression or exaggeration.

I have travelled over most of the Illinois Central Railroad, and seen the astonishing improvements to which that great enterprise has given birth. Europeans, in common with Americans, are familiar with the wonderfully liberal terms on which the Company are offering their fertile lands to actual settlers. They have made thousands of industrious families the possessors of noble homes, and will enable other thousands to become equally independent. I have given a connected history of the Company's lands, with some items of information heretofore unpublished, which will be useful both to foreign and domestic readers.

It is known that foreigners are now seeking this country in larger numbers than for several years past. This coming stream of immigration promises to expand into greater volume than ever. Multitudes of these are ignorant of our true condition, and need correct information. The majority are in search of land. Even our own citizens are deplorably ignorant of where to find the most eligible, and how to secure it. The facts contained in these pages have been collated with especial reference to the wants of both these classes of inquirers.

Some pages, not mentioned as quotations from other writers, may be recognized by the reader as having already appeared in the columns of different newspapers. All such were written by myself. Where the labors of others in the same field of inquiry have been used, the proper acknowledgment has been made.

CONTENTS.

CHAPTER I.

CHAPTER II.

CHAPTER III.

CHAPTER IV.

CHAPTER V.

CHAPTER VI.

CHAPTER VII.

CHAPTER VIII.

CHAPTER IX.

CHAPTER X.

CHAPTER XI.

CHAPTER XII.

CHAPTER XIII.

CHAPTER XIV.

HOW TO GET A FARM,

AND

WHERE TO FIND ONE.

CHAPTER I.

Poverty no Hindrance—Government Lands—Free Farms—The Homestead Law—Its Friends and Enemies—Settlers in Wisconsin—Germans in the Union—Immigration—A Southern Homestead Law—Continued Grants of Public Land.

THE buyer of a commodity seeks to purchase it at the lowest price; the seller, to dispose of it at the highest. This is the unvarying law of trade. The wealthy merchant acts up to it as closely as the poor man whose whole capital is the shilling on which he expects to dine and sup. It may be said, indeed, that it is the successful practice of this rule that constitutes the difference between the rich and the poor. It breaks down the barrier between the two, and elevates the latter to the condition of the former; for it is an accepted dogma of trade, that a thing cheaply purchased is already half sold.

Apply it to the acquisition of land. The man desirous of obtaining a farm seeks to obtain the greatest number of acres for the smallest amount of money. It is as much the governing principle of the rich as of the poor. Common sense, sharpened by long habit, teaches it to the former, but necessity teaches it to the latter. But it happens that the poor of this country cannot allege poverty as a bar to the acquisition of as much land as one man ought to possess. The vast public domain of the Union has been thrown open for them to enter in upon it as a gift. No such munificence has been displayed by any other government, either ancient or modern. When the Norman overran and conquered England, the land was partitioned off among those who assisted in the subjugation; but the mere poor man received no share because of his poverty. In our own day, the boundless fields of Australia and New Zealand are sold, not given away. This government alone has enunciated the principle that the poor man who desires to acquire land is entitled to it without price. It seeks no money compensation, but looks for remuneration to the growth and prosperity of the nation consequent on the settlement and cultivation of its vast unoccupied domain. The stranger from a foreign country, though he neither fought for it nor has been taxed for it, comes in an equal sharer with the native-born citizen.

Such lands would therefore seem to be cheaper than all others, and hence the most to be sought after. Price has no bearing upon them—they are to be given away. The causes of this unexampled liberality, the men in whose comprehensive statesmanship it originated, the opposition they encountered in its advocacy, and the conditions on which the great boon was finally wrung from its slaveholding enemies, should be fully known and understood.

It will be seen hereafter how immense the public domain yet is, even after the squandering of millions of acres on speculators and monopolists, which the last few years have witnessed. What disposition was to be made of this vast domain, was a question which long occupied the minds of thoughtful men, and of all who had the best interests of society at heart. Like most other questions in this country, it degenerated ultimately into one of party. It was clearly seen by one body of citizens that unless some radical change were made in the law, the public domain would continue to be the spoil of monopolists and speculators, the inevitable end of which would be the creation of an odious landed aristocracy. To prevent an evil so dangerous to public liberty, they determined that the only remedy was to set it aside for the exclusive use of actual settlers, in small quantities, giving it to them either at a nominal price, or as an absolute gift.

The question was an exceedingly simple one, if to be decided on its own merits. But no sooner had the free-land policy been enunciated, than the slave-power rose up in opposition. It was a measure in the interest of freedom, and slavery could not tolerate it. As the latter had for many years controlled the action of the government, so it was to override it now. Being itself a huge landed aristocracy, it saw with instant alarm the prospect of a multitude of small freeholds being established, knowing that in such a community an aristocracy could not exist. It had uniformly been hostile to pre-emption laws, and all others which tended to aid the settler in acquiring a small tract of land, and hence its bitter opposition to the free-soil scheme. Such settlers would be working men, mostly from the Free States, who would not only till the soil with their own hands, but would build school-houses, establish newspapers, and diffuse education. As no such community of intelligent toilers was permitted in the South, so should it be forbidden in the West.

On the 20th of January, 1859, a bill relating to pre-emptions being before the House of Representatives, Mr. Grow, of Pennsylvania, moved a section that no public land should thereafter be exposed to public sale by the President, unless it had been surveyed for ten or more years before such sale. The force and effect of this provision would be to give pre-emptors a start of ten years ahead of the speculators, that is, settlers would have ten years in which to choose, buy, or locate on the public lands before they could be sold to the speculators—thus giving the poor and industrious man abundant time to clear up his farm and pay for it from the productions of the soil.

The slave-power wanted no such liberty extended to the poor man. It therefore sought to defeat the bill; but Mr. Grow's amendment was adopted by a vote of 98 to 81.

The Republican vote was unanimous in its favor, and the entire slave-power voted against it, nine only excepted. Mr. Grow's amendment thus became part of the bill; but when the vote on the bill itself came to be taken, 91 Republicans voted for it, while the whole body of slaveholders, with their Northern allies, 95 in number, went against it. Only two members from the Slave States voted for the bill, Mr. Blair, of Missouri, and Mr. Winter Davis of Maryland, who represented the free-labor interests of Baltimore.

In February, the Homestead Bill was voted on in the House, and was passed by 120 to 76, only three Southern members voting for it. The bill was killed in the Senate by smothering it, all but five of the Southern Senators going against it. It was then abandoned for the session. In both Houses of Congress the Republicans had gone solid for it, while the slaveholders and their allies had so unanimously opposed it as to insure its defeat.

In 1860, another Homestead Bill was introduced into the House by the Republicans, and was passed by 115, all from the Free States but one, to 65 against it, all from the Slave States but one, and he a Pennsylvanian. When this bill went into the Senate, it was superseded by a substitute, which the House subsequently accepted, with slight amendments, the Republicans as usual voting for free homes, and the slaveholders and their allies opposing them. This took place in June. But Buchanan, then President, and the feeble and truculent tool of the slaveholders, vetoed the beneficent enactment, and once more it fell to the ground.

But undismayed by these reverses, the friends of

the bill persevered in their determination to provide free homes for the poor; and overcoming all opposition, passed the present law, which Mr. Lincoln, on the 20th of May, 1862, did not hesitate to sign. The provisions of this act are as follow:

AN ACT to Secure Homesteads to Actual Settlers on the Public Domain, and to Provide a Bounty for Soldiers in lieu of Grants of the Public lands.

Be it enacted by the Senate and House of Representatives of the United States of America in Congress assembled: That any person who is the head of a family, or who has arrived at the age of twenty-one years, and is a citizen of the United States, or who shall have filed his declaration of intention to become such, as required by the naturalization laws of the United States, and who has never borne arms against the United States Government, or given aid and comfort to its enemies, shall, from and after the 1st of January, 1863, be entitled to enter one quarter section, or a less quantity, of unappropriated public lands, upon which said person may have filed a pre-emption claim, or which may, at the time the application is made, be subject to pre-emption at $1.25, or less, per acre; or eighty acres or less of such unappropriated lands, at $2.50 per acre, to be located in a body, in conformity to the legal subdivisions of the public lands, and after the same shall have been surveyed: *Provided*, That any person owning and residing on land may, under the provisions of this act, enter other land lying contiguous to his or her said land, which shall not, with the land so already owned and occupied, exceed in the aggregate 100 acres.

SECTION 2. *And be it further enacted*, That the person applying for the benefit of this act shall, upon application to the Register of the Land-office in which he or she is about to make such entry, make affidavit before the said Register or Receiver that he or she is the head of a family, or is twenty-one years or more of age, or shall have performed service in the army of the United States, and that he has never borne arms against the Government of the United States, or given aid and comfort to its enemies, and that such application is made for his or her exclusive use and benefit, and that said entry is made for the purpose of actual settlement and cultivation, and not either directly or indirectly for the use or benefit of any other person or persons whomsoever; and upon filing the said affidavit with the Register or Receiver, and on payment of $10, he or she shall thereupon be permitted to enter the quantity of land specified: *Provided, however,* That no certificate shall be given or patent issued therefor until the expiration of five years from the date of such entry; and if, at the expiration of such time, or at any time within two years thereafter, the person making such entry—or if he be dead, his widow; or, in case of her death, his heirs or devisee; or, in case of a widow making such entry, her heirs or devisee, in case of her death—shall prove by two credible witnesses that he, she, or they have resided upon or cultivated the same for the term of five years immediately succeeding the time of filing the affidavit aforesaid, *and shall make affidavit that no part of said land has been alienated, and that he has borne true allegiance to the Government of the United States;* then, in such case, he, she, or they, if at that time a citizen of the United States, shall be entitled to a patent, as in other cases provided for by law: *And provided further,* That in case of the death of both father and mother, leaving an infant child, or children, under twenty-one years of age, the right and fee shall enure to the benefit of said infant child or children; and the executor, administrator, or guardian may, at any time within two years after the death of the surviving parent, and in accordance with the laws of the State in which such children for the time being have their domicil, sell said land for the benefit of said infants, but for no other purpose; and the purchaser shall acquire the absolute title by the purchase, and be entitled to a patent from the United States, on payment of the office fees and sum of money herein specified.

SEC. 3. *And be it further enacted*, That the Register of the Land-office shall note all such applications on the tract books and plats of his office, and keep a register of all such entries, and make return thereof to the General Land-office, together with the proof upon which they have been founded.

SEC. 4. *And be it further enacted*, That no lands acquired under the provisions of this act shall in any event become liable to the satisfaction of any debt or debts contracted prior to the issuing of the patent therefor.

SEC. 5. *And be it further enacted*, That if, at any time after the filing of the affidavit, as required in the second section of this act, and before the expiration of the five years aforesaid, it shall be proven, after due notice to the settler, to the satisfaction of the Register of the Land-office, that the person having filed such affidavit shall have actually changed his or her residence, or abandoned the said land, or shall have ceased to occupy said land for more than six months at any time, then and in that event the land so entered shall revert to the government.

SEC. 6. *And be it further enacted*, That no individual shall be permitted to acquire title to more than one quarter section under the provisions of this act; and that the Commissioner of the General Land-office is hereby required to prepare and issue such rules and regulations, consistent with this act, as shall be necessary and proper to carry its provisions into effect; and that the Registers and Receivers of the several land-offices shall be entitled to receive the same compensation for any lands entered under the provisions of this act that they are now entitled to receive when the same quality of land is entered with money, one-half to be paid by the person making the application at the time of so doing, and the other half on the issue of the certificate by the person to whom it may be issued; but this shall not be construed to enlarge the maximum of compensation now prescribed by law for any Register or Receiver: *Provided*, That nothing contained in this act shall be so construed as to impair or interfere in any manner whatever with existing pre-emption rights: *And provided, further*, That all persons who may have filed their applications for a pre-emption right prior to the passage of this act shall be entitled to all privileges of this act. *Provided further*, That no person who has served, or may hereafter serve, for a period of not less than 14 days in the army or navy of the United States, either regular or volunteer, under the laws thereof, during the existence of an actual war, domestic or foreign, shall be deprived of the benefits of this act on account of not having attained the age of 21 years.

SEC. 7. *And be it further enacted*, That the fifth section of the act entitled "An act in addition to an act more effectually to provide for the punishment of certain crimes against the United States, and for other purposes," approved the 3d of March, in the year 1857, shall extend to all oaths, affirmations, and affidavits, required or authorized by this act.

SEC. 8. *And be it further enacted*, That nothing in this act shall be so construed as to prevent any person who has availed him or herself of the benefit of the first section of this act, from paying the minimum price, or the price to which the same may have graduated, for the quantity of land so entered at any time before the expiration of the five years, and obtaining a patent therefor from the Government, as in other cases provided by law, on making proof of settlement and cultivation as provided by existing laws granting pre-emption rights.

Here is land for almost nothing. A quarter section is a hundred and sixty acres. The whole cost of obtaining such a farm is the ten dollars to be paid to the Receiver of the Land-office in which the farm may be located. On payment of this sum he enters into immediate possession, and after remaining five years upon it, he receives a patent from the government, which is equivalent to a deed in fee.

It may be supposed that this cheap way of getting a farm would occasion an instantaneous rush from East to West, to secure locations on the public domain, as well as an enormous influx of European immigrants. The act did not go into effect until January 1, 1863; yet, within four months from that date, notwithstanding the troubled state of the country, more than a million of acres were taken up under its provisions, and, by the close of September, this amount was increased to nearly a million and a half. But the great bulk of enterprising and adventurous Americans have either been drawn into the army or been too much occupied at home by the pressure of business forced upon them by the brisk demand for manufactures, occasioned by the war, to undertake the founding of a new home in the West. Neither of these classes has been at full liberty to embrace the provisions of this beneficent act. Neither, until very recently, have Europeans been well enough informed of our actual condition during the rebellion, to feel themselves safe in venturing among us, even for the purpose of securing the magnificent gift which Government holds out for their acceptance.

They have been led by rebel emissaries to believe our whole Northern and Western country to be the scene of battle, with desolation everywhere, and safety nowhere. The same dishonest agencies have been employed in leading them to believe that foreigners were conscripted at the moment of their landing among us. As men avoid rather than seek tumult, so, from these causes, the foreigner has been content to remain at home. But when the country shall have become entirely at peace, and when the provisions of the Homestead Law shall be thoroughly known in Europe, we may look with confidence for a revival of the vast stream of immigration which, a few years since, was seen pouring into our country.

What this influx has already done for us may be learned by looking at the single State of Wisconsin. The Legislature of that State found it necessary, in 1864, to order the Governor's message to be printed in eight different languages—English, German, Norwegian, Irish, Welsh, Holland, French, and Bohemian. "The North American" remarks, on this singular spectacle, that, in Wisconsin, "the old vigorous Teutonic stock is thus largely represented. The sons of the Jarls and Vikings; the descendants of Eric and Hengist; the riders of the sea, and for many generations its rulers—a brave, thrifty, intelligent, and economical people—are building up the Northwest with a rapidity which is most wonderful, and with a strength of basis which cannot be toppled over. They have contributed nobly to the nation in putting down the rebellion. They will contribute even more largely to its future welfare. They draw to them, by the irresistible magnetism of prosperity and happiness, hundreds of thousands who enjoy little of either at home; they assimilate readily with our interests and institutions. Let them continue to come. No long period will elapse, nor many generations pass to the great majority, before they will be bone of our bone and flesh of our flesh—wedded into our great unity: an element of new strength, a means of more lasting national coherence and vigor. They are of the pillars which support the power of the present and give promise to the future."

It was the cheap Government lands which drew to us all this mixed population of Wisconsin, as well as the overshadowing immigration from Germany. The still cheaper lands that can now be secured, will bring them hither in even greater numbers. The census of 1860 shows how powerful has been the attraction of cheap farms. The percentage of native Germans in this country at that period was as follows:

Wisconsin	15.97	California	7.10
Indiana	14.94	New York	6.61
Minnesota	10.59	Maryland	6.39
Illinois	7.65	Iowa	5.71
Missouri	7.50	Michigan	5.18
Ohio	7.19	New Jersey	5.03

Pennsylvania	4.74	Rhode Island	0.47
District Columbia	4.33	South Carolina	0.38
Kansas	4.03	Tennessee	0.35
Louisiana	3.48	Florida	0.34
Texas	3.40	Alabama	0.27
Kentucky	2 36	Arkansas	0.26
Oregon	2.06	Mississippi	0.25
The Territories	1.86	Georgia	0.23
Connecticut	1.85	New Hampshire	0 13
Delaware	1.13	North Carolina	0.08
Massachusetts	0.81	Vermont	0.07
Virginia	0.66	Maine	0.06

The total foreign-born population of the Union was 4,136,175, or 13.15 per cent. of the aggregate population. The English formed 1.37 per cent., the Irish 5.12, the Germans 4.14. The number of the natives of Germany was 1,301,136. The number of Germans (including their children born in this country) was four millions.

When this volume was ready for the press, the settlement of the public lands, under the provisions of the Homestead Law, was rapidly increasing. Some portions of Europe had already been made acquainted with our true condition, by means of intelligent agents sent there to circulate facts and information; while the subsequent movement in Congress in aid of immigration, attracted general attention abroad. Early in 1864, England and Ireland began to throw off their swarms of adventurers. In April, the American Consul at Liverpool wrote to Mr. Seward, as follows:

"Emigration may be said never to have been so active as it is now. It is quite unprecedented. For the past two months all the emigrant vessels from Liverpool to the States, both with steam and sails, have taken emigrants to their utmost capacity. At the present time there are not half enough ships to carry those who want to go. I called this morning on two or three of the leading shipping houses to ascertain the true state of the business, and will briefly detail what I learned. Inman's steamers—the Liverpool, New York, and Philadelphia line—told me that every passage on all their steamers up to the 18th of May next, is now engaged, and one-half of those of the steamers to sail after this period up to the 1st of June. Guion & Co., and C. Grimshaw & Co., two other large houses, told me that all the passages on their respective vessels to sail between now and the 1st day of June next, are already taken, and that they are turning off people every day for want of accommodations; that they are so pressed that they do not know what to do. They have not half vessels enough, and cannot procure them to carry the passengers that want to go. What they say will apply with equal force to all the other shippers at this port. A large proportion of the emigrants have had their passage paid in the States. These have a preference. They have raised the price of their tickets for passage, within the last few weeks, at least a third higher than they were. All the vessels sailing are filled with passengers, and the only way emigration can now be increased, so far as England and Ireland are concerned, is to increase the means of transportation. One of the houses told me this morning that they could send out fifty thousand emigrants in two months if they had the ships to carry them."

Here, then, is one way to get a farm. It is, beyond all question, the cheapest, surest, and most expeditious of any that can be suggested. It may also be the least laborious; but whether it is the most desirable in the end, each aspirant must determine for himself. It will suit many, but cannot be expected to suit all. To the strong and hardy, such as are accustomed to rough work and humble fare, it will probably be the easiest method. It must be so to thousands, or they would not so readily embrace it. How such a farm may be put in shape, and what it may be expected to produce, will be indicated in a future chapter.

Of this Homestead Law, Mr. Julian, of Indiana, thus speaks in his eloquent argument on the bill to extend its provisions to the soldiers:

"Its enactment was a long delayed but magnificent triumph of freedom and free labor over the slave-power. While that power ruled the Government its success was impossible. By recognizing the dignity of labor and the equal rights of the million, it threatened the very life of the oligarchy which had so long stood in its way. The slaveholders understood this perfectly; and hence they resisted it, reinforced by their Northern allies, with all the zeal and desperation with which they resisted abolitionism itself. Its final success is among the blessed compensations of the bloody conflict in which we are plunged. This policy takes for granted the notorious fact that our public lands have practically ceased to be a source of revenue. It recognizes the evils of land monopoly on the public domain, as well as in the old States, and looks to its settlement and improvement as the true aim and highest good of the Republic. It disowns, as iniquitous, the principle which would tax our landless poor men a dollar and a quarter per acre for the privilege of cultivating the earth; for the privilege of making it a subject of taxation, a source of national revenue, and a home for themselves and their little ones. It assumes, to use the words of General Jackson, that 'the wealth and strength of a country are its population,' and that 'the best part of that population are the cultivators of the soil.' This bold and heroic statesman urged this policy thirty-two years ago; and had it then been adopted, coupled with adequate guards against the greed of speculators, millions of landless men, who have since gone down to their graves in the weary conflict with poverty and hardship, would have been cheered and blest with independent homes on the public domain. Wealth incalculable, quarried from the mountains and wrung from the forests and prairies of the West, would have poured into the Federal coffers. The question of slavery in our national territories would have found a peaceful solution in the steady advance and sure empire of free labor, whilst slavery, in its strongholds, girdled by free institutions, might have been content to die a natural death, instead of ending its godless career in an infernal leap at the nation's throat."

In the following extract Mr. Julian foreshadows the establishment of another vast land monopoly in the South, or rather the substitution of new monopolists in place of the slaveholders, unless the operation of the Homestead Law is extended to the rebel States:

"We shall certainly win; and our triumph will inevitably divest the title to a vast body of land in the rebel States, and place it under our control. I think it entirely safe to conclude that it will constitute more than half, and probably three-fourths, of all the cultivated lands in the rebellious districts. It will certainly, in any event, cover many millions of acres. It will include all lands against which proceedings *in rem* shall be instituted, under the provisions of the act to suppress insurrections, and to punish treason and rebellion, approved July 17th, 1862; all lands which may be sold under the provisions of the act for the collection of direct taxes in insurrectionary districts, approved June 7th, 1862; and all lands which may be sold under the provisions of the act to provide internal revenue to support the Government, approved July 1st of the same year.

"What shall be done with these immense estates, brought within our power by the acts of rebels? One or two policies, radically antagonistic, must be accepted. They must be allowed to fall into the hands of speculators, and become the basis of new and frightful monopolies, or they must be placed under the jurisdiction of the Government, in trust for the people. The alternative is now presented, and presses upon us for a speedy decision. Under the laws of Congress now in force, unchecked by counter legislation, these lands will be purchased and monopolized by men who care far more for their own mercenary gains than for the real progress and glory of our country. Instead of being parcelled out into small homesteads, to be tilled by their own independent owners, they will be bought in large tracts, and thus not only deprive the great mass of landless laborers of the opportunity of acquiring homes, but place them at the mercy of the lords of the soil. The old order of things will be swept away, but a new order, scarcely less to be deplored, will succeed. In place of the slaveholding landowner of the South, lording it over hundreds of slaves and thousands of acres, we shall have the grasping monopolist of the North, whose dominion over the freedman and poor whites will be more galling than slavery itself, which in some degree tempers its despotism through the interest of the tyrant in the health and welfare of his victims. The maxim of the slaveholder that capital should own labor, will be as frightfully exemplified under the system of wages slavery, the child of land monopoly, as under the system of chattel slavery, which has so long scourged the Southern States. What we should demand is, a policy that will guarantee homes to the loyal millions who need them, and thus guard their most precious rights and interests against the remorseless exactions of capital and the pitiless rapacity of avarice."

The reading of a law so comprehensive as this will naturally induce a belief that, so far as the public domain is concerned, it is a final settlement of an angry question. But, unfortunately, this is not the fact. Mr. Julian says the overthrow of the Homestead Law is already threatened, both directly and indirectly. "Since the date of its passage," he says, "Congress has granted nearly 7,000,000 of acres for the benefit of agricultural colleges, and about 20,000,000 to aid in the construction of railroads. There are now pending before Congress (March 18, 1864), bills making other grants for railroads amounting to nearly 70,000,000 of acres. We have a project before us which grants nearly 7,000,000 of acres for the education of the children of soldiers; another, granting 200,000 acres in Michigan for the establishment of female colleges, which, of course, would be extended to the other States; and another, granting 10,000,000 of acres for the establishing of normal schools for young ladies. Every day witnesses the birth of new projects, by which our public lands may be frittered away, and the beneficent policy of the Homestead Law mutilated and destroyed."

Here are grants, perfected and in embryo, which embrace nearly 115,000,000 of acres of the lands which had been consecrated to free homes. Vast as is the quantity, the remainder is still large enough, as will be seen hereafter, for many millions of families.

CHAPTER II.

Number of Free Farms—Population, Present and Future—Increase of Public Wealth—Past and Future Immigration—Gold Mines—Farms—Enough for All.

It is known that when land could be obtained from Government at $1.25 per acre, the demand was very active, both from settlers and speculators. As the same description of lands are hereafter to be given away, many persons will presume that they will be rapidly absorbed by claimants. But there are two potent causes to prevent such result—first, the obligation to occupy the land for five years before any title whatever can be acquired, and secondly, the enormous quantity to be distributed. The following remarkable statistics on this subject are given by Mr. Samuel B. Ruggles, in his late report to the International Statistical Congress:

"The territorial area of the United States at the peace of 1783, then bounded west by the Mississippi river, was 820,680 square miles, about four times that of France, which is stated to be 207,145, exclusive of Algeria. The purchase from France of Louisiana, in 1804, added to this area 899,680 square miles. Purchases from Spain, and from Mexico, and the Oregon treaty with England, added the further quantity of 1,215,907 square miles; making the total present territory at 2,936,166 square miles, or 1,879,146,240 acres.

"Of this immense area, possessing a great variety of climate and culture, so large a portion is fertile that it has been steadily absorbed by the rapidly increased population. In May, 1863, there remained undisposed of, belonging to the Government of the United States, 964,901,625 acres.

"To prevent any confusion of boundaries, the lands are carefully surveyed and allotted by the Government, and are then granted gratuitously to actual settlers, or sold for prices not exceeding $1.25 per acre to purchasers other than settlers. It appears by the report of the Commissioner of the General Land-office, that the quantity surveyed and ready for sale in September, 1862, was 135,142,999 acres. The report also states, that the recent discoveries of rich and extensive gold fields in some of the unsurveyed portions, are rapidly filling the interior with a population whose necessities require the speedy survey and disposition of large additional tracts. The immediate survey is not, however, of vital importance, as the first occupant practically gains the pre-emptive claim to the land after the survey is completed. The cardinal, the great continental fact, so to speak, is this: that the whole of this vast body of land is freely open to gratuitous occupation, without delay or difficulty of any kind."

All these lands will necessarily rise in value as settlements are scattered through them. Our population is increasing with a rapidity not witnessed in any other country, and it is notorious that it is population which gives value to land. In 1860, we had 31,455,080 inhabitants, of whom 4,441,766 were colored, and of these, 3,953,760 were slaves. Henceforth they may be counted as freemen. The increase

of population since the establishment of the government has been as follows, as given by Mr. Ruggles:

1790	3,929,827
1800	5,305,937, increase 35.02 per cent.
1810	7,239,814, increase 36.45 per cent.
1820	9,638,191, increase 33.13 per cent.
1830	12,866,020, increase 33.49 per cent.
1840	17,069,453, increase 32.67 per cent.
1850	23,191,876, increase 35.87 per cent.
1860	31,445,080, increase 35.59 per cent.

"This rate of progress, especially since 1820, is owing in part to immigration from foreign countries.

"There arrived, in 10 years,—

From 1820 to 1830	244,490
From 1830 to 1840	552,000
From 1840 to 1850	1,558,300
From 1850 to 1860	2,707,624
Total	5,062,414

"Being a yearly average of 126,560 for the last 40 years, and 270,762 for the last ten years."

The rebellion checked the tide of foreign immigration; but in 1863 it again commenced setting towards our shores. Mr. Ruggles says:

"The records of the Commissioners of Emigration of New York show that the arrivals at that port alone have been, for

	From Ireland.	From Germany.	Total, including all other countries.
1861	27,754	27,159	65,529
1862	32,217	27,740	76,306
1863, up to Aug. 20, 7⅔ mos.	64,465	18,724	about 98,000

"The proportions of the whole number of 5,062,414 arriving from foreign countries in the forty years from 1820 to 1860, were as follows:

From Ireland...........................	967,366	
From England..........................	302,665	
From Scotland..........................	47,800	
From Wales............................	7,935	
From Great Britain and Ireland............	1,425,018—	2,750,784
From Germany.........................	1,546,976	
From Sweden..........................	36,129	
From Denmark and Norway..............	5,540—	1,588,145
From France...........................	208,063	
From Italy............................	11,302	
From Switzerland......................	37,732	
From Spain............................	16,245	
From British America..................	117,142	
From China (in California almost exclusively)	41,443	
From all other countries, or unknown.......	291,558—	723,485
Total...........		5,062,414

"It is not ascertainable how many have returned to foreign countries, but they probably do not exceed a million. If the present partial check to immigration should continue, though it is hardly probable, the number of immigrants for the decade ending in 1870 may possibly be reduced from 2,707,624 to 1,500,000.

"The ascertained average increase of the whole population in the seven decades from 1790 to 1860, which is very nearly 33⅓ per cent., or one-third for each decade, would carry the present numbers (31,445,080)

By the year 1870, to..........................	41,926,750
From which deduct for the possible diminution of immigrants, as above........................	1,207,624
There would remain......................	40,719,126

"Mr. Kennedy, the experienced Superintendent of the census, in the Compend published in 1862, at page 7, estimates the population of 1870 at 42,318,432, and of 1880, at 56,450,241. The rate of progress of the population of the United States has much exceeded that of any of the European nations. The experieneed statisticians in the present Congress can readily furnish the figures precisely showing the comparative rate.

The population of France was, in

1801	27,349,003	1841	34,230,178
1821	30,461,875	1851	35,283,170
1831	32,569,223	1861	37,472,132

"Being about 37 per cent. in the 60 years. It does not include Algeria, which has a European population of 192,746.

"The population of Prussia has increased since 1816, as follows:

1816	10,319,993	1849	16,296,483
1822	11,664,133	1858	17,672,609
1834	13,038,970	1861	18,491,220
1840	14,928,503		

"Being at the rate of 79 per cent. in 45 years.

"The population of England and Wales was, in

1801	9,156,171	1841	16,035,198
1811	10,454,529	1851	18,054,170
1821	12,172,664	1861	20,227,746
1831	14,051,986		

"Showing an increase of 121 per cent. in 60 years, against an increase in the United States in 60 years, of 593 per cent.

"The natural and inevitable result of this great increase of population, enjoying an ample supply of fertile land, is seen in a corresponding advance in the material wealth of the people of the United States. For the purpose of State taxation, the values of their real and personal property are yearly assessed by officers appointed by the States. The assessment does not include large amounts of property held by religious, educational, charitable, and other associations, exempted by law from taxation, nor any public property of any description. In actual practice, the real property is rarely assessed for more than two-thirds of its cash value, while large amounts of personal property, being easily concealed, escape assessment altogether.

"The assessed value of that portion of property which is thus actually taxed increased as follows: In 1791 (estimated), $750,000,000; 1816 (estimated), $1,800,000,000; 1850 (official valuation), $7,135,780,228; 1860 (official valuation), $16,159,616,068, showing an increase in the last decade alone of $9,023,835,840.

"A question has been raised, in some quarters, as to the correctness of these valuations of 1850 and 1860, in embracing in the valuation of 1850, $961,000,000, and in the valuation of 1860, $1,936,000,000, as the assessed value of slaves, insisting that black men are persons and not property, and should be regarded, like other men, only as producers and consumers. If this view of the subject should be admitted, the valuation of 1850 would be reduced to $6,174,780,000, and that of 1860, to $14,223,618,068, leaving the increase in the decade $8,848,825,840.

"The advance, even if reduced to $8,048,825,840, is sufficiently large to require the most attentive examination. It is an increase of property over the valuation of 1850, of 130 per cent., while the increase of population in the same decade was but 35.99 per cent. In seeking for the cause of this discrepancy, we shall reach a fundamental and all-important fact, which will furnish the key to the past and to the future progress of the United States. It is the power they possess, by means of canals and railways, to practically abolish the distance between the seaboard and the wide-spread and fertile regions of the interior, thereby removing the clog on their agricultural industry, and virtually placing them side by side with the communities on the Atlantic. During the decade ending in 1860, the sum of

$413,541,510 was expended within the limits of the interior central group known as the 'food-exporting States,' in constructing 11,212 miles of railway to connect them with the seaboard. The traffic receipts from those roads were: In 1860, $31,335,031; in 1861, $35,305,509; in 1862, $44,908,405.

"The saving to the communities themselves in the transportation, for which they thus paid $44,908,405, was at least five times that amount; while the increase in the exports from that portion of the Union greatly animated not only the commerce of the Atlantic States, carrying those exports over their railways to the seaboard, but the manufacturing industry of the Eastern States, that exchange the fabrics of their workshops for the food of the interior.

"By carefully analyzing the $8,048,825,840 in question, we find that the six manufacturing States of New England received $735,754,244 of the amount; that the middle Atlantic, or carrying and commercial States, from New York to Maryland, inclusive, received $1,834,911,579; and that the food-producing interior itself, embracing the eight great States of Ohio, Indiana, Illinois, Michigan, Wisconsin, Minnesota, Iowa, and Missouri, received $2,810,000,000. This very large accession of wealth to this single group of States is sufficiently important to be stated more in detail. The group, taken as a whole, extends from the western boundaries of New York and Pennsylvania to the Missouri river, through 14 degrees of longitude, and from the Ohio river north to the British dominions, through 12 degrees of latitude. It embraces an area of 441,167 square miles, or 282,134,688 acres, nearly all of which is arable and exceedingly fertile, much of it in prairie and ready at once for the plough. There may be a small portion, adjacent to Lake Superior, unfit for cultivation, but it is abundantly compensated by its rich deposits of copper and of iron of the best quality.

"Into this immense natural garden, in a salubrious and desirable portion of the temperate zone, the swelling stream of population, from the older Atlantic States and from Europe, had steadily flowed during the last decade, increasing its previous population from 5,403,595 to 8,957,690; an accession of 3,554,095 inhabitants, gained by the peaceful conquest of nature, fully equal to the population of Silesia, which cost Frederick the Great the seven years' war, and exceeding that of Scotland, the subject of struggle for centuries.

"The rapid influx of population into this group of States increased the quantity of the 'improved' land, thereby meaning farms more or less cultivated, within their limits, from 26,680,361 acres, in 1850, to 51,826,395, in 1860; but leaving a residue, yet to be improved, of 230,308,293 acres. The area of 25,146,054 acres, thus taken in ten years from the prairie and the forest, is equal to seven-eighths of the arable area of England, stated by its political economists to be 28,000,000 of acres.

"The area embraced in the residue will permit a similar operation to be repeated eight times successively, plainly demonstrating the capacity of this group of States to expand their present population of 8,957,690, to at least 30,000,000, if not 40,000,000 of inhabitants, without inconvenience.

"The effects of this influx of population in increasing the pecuniary wealth as well as the agricultural products of the States in question, are signally manifest in the census. The assessed value of their real and personal property ascended from $1,116,000,000, in 1850, to $3,926,000,000, in 1860, showing a clear increase of $2,810,000,000. We can best measure this rapid and enormous accession of wealth, by comparing it with an object which all nations value—the commercial marine. The commercial tonnage of the United States, in 1840, was, 2,180,764 tons; in 1850, 3,535,454 tons; in 1860, 5,358,808 tons.

"At $50 per ton, which is a full estimate, the whole pecuniary value of the 5,358,808 tons, embracing all our commercial fleets on the oceans, and the lakes, and the rivers, numbering nearly thirty thousand vessels, would be but $267,940,000; whereas the increase in the pecuniary value of the States under consideration, in each year of the last decade, was $281,000,000. Five years' increase would purchase every commercial vessel in the Christian world.

"But the census discloses another very important feature, in respect to these interior States, of far higher interest to the statisticians, and especially to the statesmen of Europe, than any which has yet been noticed, in their vast and rapidly increasing capacity to supply food, both vegetable and animal, cheaply and abundantly, to the increasing millions of the Old World. In the last decade their cereal products increased from 309,950,295 bushels, to 558,160,323 bushels, considerably exceeding the whole cereal product of England, and nearly if not quite equal to that of France. In the same period the swine, who play a very important part in consuming the large surplus of Indian corn, increased in number from 8,536,182 to 11,039,352, and the cattle from 4,373,712, to 7,204,810. Thanks to steam and the railway, the herds of cattle who feed on the meadows of the Upper Mississippi are now carried in four days, through eighteen degrees of longitude, to the slaughter-houses on the Atlantic.

"It is difficult to furnish any visible or adequate measure for a mass of cereals so enormous as 558,000,000 of bushels. About one-fifth of the whole descends the chain of lakes, on which 1,300 vessels are constantly employed in the season of navigation. About one-seventh of the whole finds its way to the ocean through the Erie canal, which has already been once enlarged for the purpose of passing vessels of two hundred tons, and is now under survey by the

State of New York for a second enlargement, to pass vessels of five hundred tons. The vessels called 'canal boats,' now navigating the canal, exceed five thousand in number, and if placed in a line, would be more than eighty miles in length.

"The barrels of wheat and flour alone, carried by the canal to the Hudson river, were, in 1842, 1,146,292; in 1852, 3,937,366; in 1862, 7,516,397.

"A similar enlargement is also proposed for the canal connecting Lake Michigan with the Mississippi river. When both the works are completed, a barrel of flour can be carried from St. Louis to New York, nearly half across the continent, for fifty cents; or a ton, from the Iron Mountain of Missouri, for $5. The moderate portion of the cereals that descends the lakes, if placed in barrels of five bushels each, and side by side, would form a line of five thousand miles long. The whole crop, if placed in barrels, would encircle the globe. Such is its present magnitude. We leave it to statistical science to discern and fully estimate the future. One result is, at all events, apparent. A general famine is now impossible; for America, if necessary, can feed Europe for centuries to come. Let the statesman and philanthropist ponder well the magnitude of the fact, and all its far-reaching consequences—political, social, and moral—in the increased industry, the increased happiness, and the assured peace of the world.

"The great metalliferous region of the American Union is found between the Missouri river and the Pacific Ocean. This grand division of the Republic embraces little more than half of its whole continental breadth. Portland, in Maine, is the meridian 70° west from Greenwich; Leavenworth, on the Missouri river, in 95°; and San Francisco, on the Pacific, in 123°. By these continental landmarks, the western or metalliferous section is found to embrace 28°, and the eastern division between the Missouri and the Atlantic, at Portland, 25° of our total territorial breadth of 53° of longitude.

"It has been the principal work and office of the American people, since the foundation of their government, to carry the machinery of civilization westward from the Atlantic to the Missouri, the great confluent of the Mississippi. So far as the means of rapid intercommunication are concerned, the work may be said to be accomplished, for a locomotive engine can now run without interruption from Portland to the Missouri, striking it at St. Joseph, just below the fortieth parallel of latitude. In the twenty years preceding 1860, a network of railways 31,196 miles in length was constructed, having the terminus of the most western link on the Missouri river. The total cost was $1,151,560,829, of which $850,900,681 was expended in the decade between 1850 and 1860. The American Government and people had become aware of the great pecuniary, commercial, and political results of connecting the ocean with the food-producing interior by adequate steam communications. But the higher and more difficult problem was then presented of repeating the effort on a scale still more grand and continental; of winning victories still more arduous over nature; of encountering and subduing the massive mountain ranges interposed by the prolongation of the Cordilleras of our sister continent through the centre of North America, rising, even at their lowest points of depression, far above the highest peaks of the Atlantic States.

"The Government, feeling the vital, national importance of closely connecting the States of the Atlantic and of the Mississippi with the Pacific with all practicable dispatch, has vigorously exerted its power. On the 1st of July, 1862, nearly fifteen months after the outbreak of the existing insurrection, and notwithstanding the necessity of calling into the field more than half a million of men to enforce the national authority, Congress passed an act for incorporating 'The Union Pacific Railway Company,' and appropriated $66,000,000 in the bonds of the United States, with large grants of land, to aid the work, directing it to be commenced at the 100th meridian of longitude, but with four branches extending to the Missouri river. The necessary surveys across the mountain ranges are now in active progress, and the construction of the eastern division, leading westward from the mouth of the Kansas river, or the Missouri, has actually commenced. The whole of that division, including that part of the line west of the 100th meridian to the foot of the Rocky Mountains, is on a nearly level plain, and is singularly easy of construction. Its western end will strike the most prominent point of the auriferous regions in the Territory of Colorado, where the annual product of gold, as stated in the official message of the Territorial governor, is from $5,000,000 to $10,000,000. The gold is there extracted by crushing-machines from the quartz, in which it is found extensively distributed, needing only the railway from Missouri to cheaply carry the necessary miners, with their machinery and supplies. The distance to that point will be about six hundred and fifty miles, which will be passed in twenty-eight hours. When completed, as it easily may be, within the next three years, it will open the way for such an exodus of miners as the country has not seen since the first discoveries in California, to which the American people rushed with such avidity, many of them circumnavigating Cape Horn to reach the scene of attraction.

"Meanwhile a corresponding movement has commenced on the Pacific, in vigorously prosecuting the construction of the railway eastward from the coast at or near San Francisco, which will cross the Sierra Nevada at an elevation of about 7,000 feet, on the eastern line of California, in the

120th parallel of longitude, and there descend into the territory of Nevada, at the rich silver mines of Washoe.

"It is not yet possible to estimate with any accuracy the extent of these deposits of gold and silver, but they are already known to exist at very numerous localities in and between the Rocky Mountains and the Sierra Nevada, not to mention the rich quartz-mining regions in California itself, which continue to pour out their volumes of gold to affect, whether for good or ill, the financial condition of the civilized world. During the last six months gold has been obtained in such quantities, from the sands of the Snake river and other confluents of the Columbia river, as to attract more than 20,000 persons to that remote portion of our metalliferous interior. The products of those streams alone for the present year are estimated at $20,000,000.

"The Commissioner of the General Land-office, in his official report of the 29th December, 1862, states as follows:

"'The great auriferous region of the United States, in the western portion of the Continent, stretches from the 49th degree of north latitude and Puget Sound, to the 30° 30′ parallel, and from the 102d degree of longitude west of Greenwich, to the Pacific Ocean, embracing portions of Dakota, Nebraska, Colorado, all of New Mexico, with Arizona, Utah, Nevada, California, Oregon, and Washington Territories. It may be designated as comprising 17 degrees of latitude, or a breadth of 1,100 miles from north to south, and of nearly equal longitudinal extension, making an area of more than a million square miles.

"'This vast region is traversed from north to south, first, on the Pacific side, by the Sierra Nevada and Cascade Mountains, then by the Blue and Humboldt; on the east, by the double ranges of the Rocky Mountains, embracing the Wahsatch and Wind River chain, and the Sierra Madre, stretching longitudinally and in lateral spurs, crossed and linked together by intervening ridges, connecting the whole system by five principal ranges, dividing the country into an equal number of basins, each being nearly surrounded by mountains and watered by mountain streams and snows, thereby interspersing this immense territory with bodies of agricultural lands equal to the support not only of miners, but of a dense population.

"'These mountains,' he continues, 'are literally stocked with minerals; gold and silver being interspersed in profusion over this immense surface, and daily brought to light by new discoveries. In addition to the deposits of gold and silver, various sections of the whole region are rich in precious stones, marble, gypsum, salt, tin, quicksilver, asphaltum, coal, iron, copper, lead, mineral and medicinal, thermal and cold springs and streams.

"'The yield of the precious metals alone of this region will not fall below one hundred millions of dollars the present year, and it will augment with the increase of population for centuries to come. Within ten years the annual product of these mines will reach two hundred millions of dollars in the precious metals, and in coal, iron, tin, lead, quicksilver, and copper, half that sum.' He proposes to subject these minerals to a government tax of eight per cent., and counts upon a revenue from this source of $25,000,000 per annum almost immediately, and upon a proportionate increase in the future. He adds, that 'with an amount of labor relatively equal to that expended in California applied to the gold fields already known to exist outside of that State, the production of this year, including that of California, would exceed four hundred millions. In a word,' says he, 'the value of these mines is absolutely incalculable.'"

The foregoing facts and deductions set forth not

only the inexhaustible quantity of land now freely open to all who choose to occupy it, but refer to its variety of character as adapted to suit the diversified wishes of the many who seek to acquire farms. The farmer can be accommodated with woodland or prairie, the lumberman or mechanic with densely wooded forest, the miller or manufacturer with mill-sites, the miner with either silver, gold, or coal.

The quantity is without limit, and the uses to which it may be profitably applied are so numerous that the most fastidious applicant may be supplied with what he wishes. Millions of families may thus obtain farms before the quantity now open for selection can be appropriated. It will require centuries to fill it up. Hence those either here or abroad, who learn for the first time that farms may be had on the simple condition of living on them for five years, may entertain no fear that a sudden absorption will deprive them of the opportunity of obtaining one. It is the monopolists and speculators who are repudiated, not the actual settler.

How this national liberality is to affect the value of land generally, may be inferred from what has followed the abolition of serfdom in Russia. That great measure threw open millions of acres to the occupancy and ownership of a people who had heretofore only tilled them for the benefit of a master. The privilege of obtaining land, even by paying for it, revolutioned the feelings and industry of the entire mass. Emancipation was completely triumphant in every respect. All the forebodings of the reactionaries have been disappointed. A recent traveller says:

"There has been no bloodshed, no excess, no social disorder, no decline of industry. Twenty-three millions of people have been raised at once from the degradation of chattelism to the dignity of freemen, by the fiat of one man, in the space of two years, in the face of a most formidable opposition of nearly the whole Russian nobility. The bitterest opponents now admit that as the operation had to be performed some time, it was well to do it at once. Intellectual and social energies which had been frozen up for centuries, are set free; the peasantry are a promising race of people, and they know how to appreciate the boon of liberty. Among the first financial results is the general rise in the price of land all through Russia, at least a million of serfs having already purchased the land which they formerly cultivated for a master. The Government systematically loans money for this object, and all the money which was formerly hidden in earthen pots is brought out and invested in land. Every peasant feels a new incentive to industry and economy, that he may be able to buy land. More houses are now built in a year than used to be built in half a dozen years. The new wants of the people give a surprising impulse to trade. The nobility, who used to spend their incomes in Paris or in Germany, are coming to live on their estates, and spend their lives in seeking to promote the improvement of the people. The appraised value of property in the kingdom is already enhanced almost beyond computation.

"The educational and religious efforts are equally signal. Already eight thousand schools have sprung into existence among the peasants, by their own efforts, aided by friends, the Government having no hand in it. Two years ago such a thing as a day-school among the peasantry was hardly known. There is great anxiety to be able to read the laws, as well as to read the Scriptures. To meet a pressing demand, the Church authorities have published the Russian New Testament at the low price of sixpence a copy.

"The changes which have already been made in the municipal arrangements of the country are equally wonderful. Within the last two years the cities of Moscow and St. Petersburg have for the first time had mayors elected by the citizens. In the peasant villages, the chief is elected by the people, and all measures are debated and settled in village meetings—the training-schools of freedom, as every philosophical observer considers our American town-meetings. An honorary local magistracy has been created all over the empire, of men of character and standing, who can execute justice between man and man, repress crime, and protect the weak against the strong."

The benefits to be conferred on this country by the Homestead Law are strikingly illustrated by the events of the slaveholders' rebellion. It has been seen that cheap lands have induced a vast immigration, and that by help of this immigration the republic has sprung, in a single lifetime, to the status of a powerful nation. Of the whole number of arrivals, ninety-five per cent. have settled in the Free States, and only five per cent. in the Slave States. An anonymous essayist presents the following views in relation to this part of the question:

"It is from the armies, raised from the former and their descendants, that the Government has been mainly enabled to overcome the rebellion. They gave to the nation its magnitude, and that magnitude alone has saved us from foreign intervention. Had the bloody ordeal fallen on us when possessed of but one-tenth of our present population, there can be little question that the intense hostility with which we are regarded by the ruling classes in the nations of Western Europe, would have dictated armed intervention, the forcible opening of the blockade, and, finally, the dismemberment of the Republic. If the magnitude of our resources and the numbers of our armies appalled our enemies, both at home and abroad, it must be borne in mind that these were but results made possible by our vast population."

"Our foes shrank from a contest with a nation which, even in the midst of an unexampled rebellion, was still able to pour its armies into the field by the million, and to sustain the Government by an incalculable store of riches. Our vast northern and western population has saved it from overthrow. If, with this great preponderance of numbers, we have found it so difficult to overcome rebellion, it will be at once perceived, that, if our population had been no greater than that of the South, the task of suppressing it would have been a sheer impossibility. Instead of literally overrunning the South, and crushing it beneath the mere weight of numbers, we should have found ourselves engaged in a war ruinously protracted, the end of which, in all human probability, would have been a destruction of the Republic."

Thus all that is dear to us as a united people, has depended on a question of numbers. The consideration of this fact may not have been embraced in the calculations of those who, many years ago, put the public lands in market at a low price; but it became a controlling element of the policy which enacted the Homestead Law. As the cheap lands have once saved the nation from destruction, so the still cheaper ones may be relied on to insure its preservation.

A recent anonymous writer on this subject furnishes the following appropriate suggestions:

"The lands given away will be worth far more to the country, peopled with an industrious population, than lying waste as they now do. They will soon yield up their treasures of grain or of cotton and tobacco to be exported, and to buy goods that will pay a duty to the Government. Peopled, they will furnish soldiers for the army, and taxes to pay their expenses, should the country need them. Before this law was passed, lands were so cheap that every man of real energy and industry could obtain a homestead if he tried, provided he could raise the means to get on to the land. This will be the chief difficulty now. Hundreds and thousands of families, to whom the land would be a priceless boon, will never be able to reach it. They have little forecast, are poor and in debt, and pretty much discouraged. They cannot find constant employment, and do not know how to employ themselves profitably. If associations could be formed for settling these lands in part by such families, it would meet the difficulty. It would help them without damaging the success of the new settlement. It would secure to them at once homesteads and full employment, which they so much need.

"Many questions are asked concerning this new law by those who desire to avail themselves of its advantages. A careful reading of the law will answer many of them. The lands are to be found in Michigan, Wisconsin, Iowa, Minnesota, Missouri, Kansas, Nebraska, Texas, and on the Pacific, in large extent, and some still in Ohio, Indiana, and Illinois, though they are probably not of a very inviting character. The lands lying along railroads are of double price, and, on account of the proximity to market, are perhaps cheaper at that rate. Only eighty acres of these can be taken by one individual.

"The old pre-emption laws are still in force, and a man may locate his land, holding by these laws until the 1st of January, when he can hold by the new law. There are land-offices in the vicinity of all these public lands, where the applicant can make known his wants and secure his homestead. It will be seen that the matter involves either the expense of a personal visit, or that of a delegate, which is a serious obstacle to the poor. The best thing that can be done, probably, in all cases by those who wish to avail themselves of this law, will be to form an association for the settlement of a township, say a hundred families or more, and send out an agent to examine and locate the lands in a body. The advantages of planting a whole Christian community in the wilderness at once, over private emigration, are too apparent to need mention here.

"A farm for ten dollars is only the raw material of a home. Houses, barns, fences, roads, bridges, churches, school-houses, and other public buildings, are to be provided after the colony is located, and these things bring heavy taxes upon every individual for a dozen years or more. A man getting a living at the East should think twice before he goes into the wilderness. It is young men just married, or about to be, men with large families and scanty means of living, and professional men with small fields of labor, that can take this step with the best prospects."

CHAPTER III.

What makes Land valuable — Prices balancing each other — How poor Men pay for high-priced Farms—A practical Illustration—A Farm for the Right Man.

While it is thus seen that there are millions of families who desire no better homestead than such as can be secured by settlement on the public domain, it is well known that there are other millions who prefer remaining in the neighborhood in which they were born. They prefer hard work there to hard work in the West. That region is new, and large portions of it are comparatively unsettled. The other is old, and possesses all the conveniences and comforts of a long-established civilization. Relations and friends are there concentrated, and among them they prefer remaining. It furnishes a quick market for all productions of the earth, and at better prices. Fruits and vegetables, which, on a thousand prairie farms, would find no purchaser, are here salable in every town or city. Here the consumers are collected in great crowded marts, while there they have not yet had time to congregate in equal masses.

Land within the seaboard region is consequently more valuable, and, as a general rule, is unattainable by small capitalists in proportion to its value. But its ability to yield quick and certain returns makes its possession extremely desirable. Its money-producing power is enormous, because of its nearness to a dense population of consumers. As to this fact it owes its chief value, so, from the same fact, the small capitalist who becomes possessed of it is enabled to pay for it by the ready and profitable market he finds for all that it may produce.

Thus price has its compensations. If the cost of land be high, the value which its productions command in the market is generally in exact proportion. High price for land, and low price for products, would be ruinous to the farmer. But let the latter maintain a just relation to the former, and if the land be skilfully worked with distinct reference to the most profitable crops it can be made to yield, the lapse of a few years will enable the industrious owner to make full payment. Wheat may be grown with profit on a prairie farm which the owner obtained as a gift, because for that grain there is a cash market at the nearest railroad station. But asparagus and cabbages would perish on the grower's hands. Wheat can be shipped to Europe, and hence its universal salability; but the vegetables must find purchasers within short distances of the spot where they were grown. So, on the other hand, the man who cultivates high-priced land within the suburbs of a great city, will lose money by raising wheat, while by cultivating asparagus and cabbages he will be certain to grow rich. The West can undersell him in wheat, but cannot compete with him in vegetables. Hence the proper adaptation of crop to location is absolutely indispensable to success.

It has been shown how the poor man can gratuitously obtain a farm where he may not happen to be desirous of locating. It remains to be shown how he can get one where he does desire to settle. To promote this laudable ambition of those whose whole capital is industry and labor, much has been already written by ingenious and generous men. Their views and plans have been different, as well as numberless. It is remarkable, however, that while some of them propose methods which would require a lifetime to make successful, none of them present difficulties too great to be in some way overcome. I refer now, as well as throughout these pages, to the man who is sober, industrious, ambitious of success, saving, and possessed of ordinary intelligence. The poverty of such may be an inconvenience, but it is no insuperable bar to progress. The men whose characters are the reverse, I do not write for. It is they who, instead of acquiring farms, invariably lose them. It will also be seen that feeble health need be no fatal discouragement, and that some men have succeeded even when comparatively disabled by incurable bodily infirmity.

A practical farmer, writing in the Albany *Country Gentleman*, in 1862, gives the following as his method of getting a farm with no cash capital to begin with. His article is in reply to a writer in the same paper, who wishes to know how to get a farm without money or capital at the outset, and who says that there are, no doubt, many men in our country who commenced life under similar circumstances, who have risen to be successful and independent farmers. The reply bears the signature of "F.," and is as follows:

"Having commenced life under circumstances substantially the same as those described by your correspondent, and having thought much on this subject, and no answer having appeared as yet, I have concluded to try and see what I can do towards helping him, and others in similar circumstances, in their laudable efforts to get a farm.

"Well do I remember the intense thought and study with which I first turned my attention to farming as a means of getting a living. Having failed in other business, for want of the capital without which I had always supposed I could not succeed in farming, I was casting about and considering what to try next, when for the first time I came across some agricultural publications, which were read with all the interest of an exciting romance, and which at once led to a determination to make farming the business of my life. But here I was met by the same difficulty as your correspondent. I had no land, nor nothing to buy with. I was in a strange country, with no friends to assist me in beginning, except such as by industry, economy, and fair dealing, I was able to make. Yet I have succeeded so far, beyond my most sanguine expectations; while my farming prospects are not only improving every year, but they are better now than ever before.

"But in answering your correspondent, I do not propose to go into the details of my own experience, but rather, as briefly as may be, try to point out the best course for young men to pursue, in order to succeed in getting farms. I am led to take this course, not only by the reluctance felt by most men of laying their private business affairs before the public, but because, in giving the combined results of reading, observation, and experience, I believe I shall be able to more effectually assist G. B. S. and others in similar circumstances.

"One of the greatest difficulties encountered by young men, in trying to get a farm, is to get a start, or in other words to get the first $500. Almost any young man that will go to work, and earn and lay up that amount of money, may, with good habits, and industry, and economy, be sure of sooner or later owning a good farm. It cannot be too strongly impressed on the minds of all young men, that the great starting point in their fortune, is to earn and lay up the first $500 or $1000. Not only for the help that amount will be in gaining more money, but in firmly fixing in their minds the principles of industry, economy, and self-denial, which are to be the foundation of their future success.

"The most usual course taken by farmers' sons to get this start, is by working out by the month for farmers; and perhaps it is the best course open to thousands of young men in our country. But a large portion of these young men are only able to get work for seven or eight months of the busy season, leaving them idle during the winter and a part of the spring and fall. The wages they will earn in this way, will not enable them to lay up money very fast. Hence the enterprising young man that is determined to succeed, will either be sure to hire out by the year, or teach school through the winter, or find some kind of job-work, by which he will be able to make good wages all of the time. He will also keep in mind, that by continual faithfulness, care, and attention, to the business of his employer, he will not only be earning and getting much higher wages than others, but he will be forming habits of care and attention, that will be highly useful as long as he lives.

By taking this course, a young man ought to lay up $100 a year, and many will lay up more. We will suppose he commences when he is 21, and when he is 25 has saved $500.

"Now, what is the best course for a young man that has earned, or by some other means come in possession of $500, who wishes to get a good farm? He desires, and ought to have one worth $4,000 or $5,000, or more, and with him the very important question is, what is the best course for him to take to get it? Now, without taking into consideration the question of going into a new country, in pursuit of cheap land, I conceive he must choose one of the following courses: He will either continue to work out, or he will take or rent a farm, or he will buy and commence farming on a small farm. It being necessary, in order for any one to pursue either one of the two last courses, to have some capital, it is not considered that there is the same necessity for working out after a man has $500 that there was before. Consequently, he may now be considered as fairly in a condition to take his choice between the three courses here pointed out. And, as undoubtedly there are many, in different parts of the country, that may find it desirable to follow each of these different ways, and many more desiring all the information they can get, in regard to the best course to pursue, perhaps it will be best to bestow some attention on each of these ways to get a farm.

"First, in regard to working out. This is a very simple, plain, straight-forward way to get a farm. It is only a continuation of the course already pointed out for those who have to start with nothing for an indefinite period of time, which will be longer or shorter in proportion to the amount desired to commence with. The advantages of this course are presented in a very favorable light by Hon. J. W. Colburn. He says:

"'Now let us for a moment look at the matter, and see what the real obstacles are which are to be overcome by the resolute young man of 21 years of age, who says, 'I will own a good farm.' His father has had the benefit of his labor up to this time, and is unable or unwilling to give him any thing to start in life. His whole capital consists in muscular strength, good health, good will, self-reliance, and correct principles. He takes the best wages he can get of a responsible farmer in the neighborhood—say $15 per month for the year, board and washing included. He pursues this course for seven years; his economy has taught him that $60 per year is sufficient for clothing and other expenses, leaving $120 at the end of each year to put at interest. At the age of 28 years he has earned and put at interest $840. What the several annual interests have been I will not stop to enumerate. It is sufficient to say that he has a sum sufficient to start him handsomely in a new country, with a half section, 320 acres, paid for, and means enough left for an outfit to commence successful operations upon his new farm. In 22 years more, with ordinary good luck, how will he be likely to stand? He is now 50 years old, and a man of wealth, and probably of character and influence.

"'Or take another view of it. Suppose at the age of 28 he should say, I don't think much of this emigrating, there is some risk in the change of climate; I like my old associations; my friends are here, my home scenes are dear to me; the girl of my choice is unwilling to go to the far west, or into a new country; I will settle in my own neighborhood. He buys an improved farm with fences and comfortable buildings, say 100 acres, at $20 per acre—pays one-half down, balance in ten years, interest annually. What will now be his condition at 50 years of age? Perhaps not as wealthy as in the first case, with equally good luck in both, for the advance of his land in value above cost in the former case would have been a little fortune; but he has made a sure thing of it as it is, has lived healthily, and saved a competency for all future wants.'

"Mr. Colburn further states that 'this is not an overdrawn picture. It is what has transpired within the observation of the writer, and at a time too when farm wages were lower than the prices here named, and much lower than at the present time.'

"Now it will be admitted on all sides that these extracts place this part of the subject in the most favorable light; that in fact it would seem that there can scarcely be any need of, even if there is any room for, saying any thing more on this side of the question; therefore it only remains to briefly allude to some of the objections that young men may find to pursuing the course so favorably presented.

"The first objection will be in regard to wages. It will be said, and with a great deal of truth, that such wages are a good deal higher than young men that work on a farm are generally able to realize. And to this it will be added that in most of the older settled sections of the country $800 or $1,000 goes but a little way towards paying for a good farm. Consequently, it will be said a young man will have to work out a great deal more than seven years, in most cases from twice to three times that length of time, before he can even pay half down for a good farm, to say nothing of the money that will be needed to begin farming with.

"Perhaps there is nothing that a spirited, enterprising young man would view with greater reluctance than the proposition for him to make up his mind to work out for from ten to twenty years of the best of his life in order to get the requisite capital to commence the business of farming with. He will probably say that he has not so much objection to working out a few years in order to get a few hundred dollars to start with. But as to working out that length of time, it is useless to talk about it; he is not going to do it. Point him to some one that has succeeded in this way; he will admit it, but say this is an exception, not the general rule, and will point to many men that have failed of ever getting farms by working out, and will say that no man with a growing family on his hands can lay up any thing, to say nothing of saving enough to buy a farm; while there are few young men that have not formed ties and made arrangements, that are not to be put off for a very indefinite period. Hence, put the case in as strong a light as we may, or argue it ever so strongly, it will be of little use. Consequently, those that would persuade young men to stick to farming, and undertake to point out a way whereby they may get a farm, will, in most cases, have to show them some other way besides working out. Yet it cannot be denied that young men do not always sufficiently appreciate this way of getting a start in the world; that in many cases it is the best thing a single man can do as long as he remains single, and that many that have left it for other business would have done better if they had not made the change.

"But we must pass on to consider the second course for a young man to take, in order to get a farm. Renting a farm, or taking one on shares, is, next to owning one, what seems to suit young men the best. G. B. S. seems to have had this course in his mind; but says that 'renting a farm, the way it is done in this part of the country, is not very desirable. It generally goes on the "skinning process," making it profitless to both parties.' Now here is the main difficulty, not only as to those taking farms, in finding this a good 'way to get a farm,' but with those having farms to let. It is a fact well understood on all hands, that, as poor business as taking or renting land may be, there are many, in all parts of the older sections of the country, that would be glad to get farms to work if they could. In this section, whenever there is a good farm to be let, there are sure to be from eight to ten, and sometimes a score or more applications for it. While at the same time there are many that would like to let their farms, were it not for this one difficulty—they are sure to be worked on the skinning system; which, while it gives them but little present profit, is injuring, if not ruining their land. Hence, I would impress on the minds of all, young or old, in the strongest language, and in the most earnest manner, the great mistake they make in thinking that, because they are working another man's farm, they cannot afford to farm well; that they are taking a course that not only gives them but little present profit, but one that, more than perhaps any thing else, tends to deprive them of the chance of getting what little they do have; and that not only will they realize a much greater present profit, by as good farming as the circumstances will allow, but should it be the case that at first they are not able to get a good farm, but have to take up with rather inferior, or badly worn land, they may be sure that, if they are doing their best, it will be known and observed, and they will have no trouble in getting better land when they wish to make a change.

"The advantages of this course may be made still plainer by taking a case, many of which may be found in all of the older sections of the country. The owner of a 100-acre farm, that has not only made the principal part of his property out of his farm, but has brought up, educated, and given his children more or less of a start in life, and who has found, by experience, that one-half of the produce of his land will support him satisfactorily, wishes to let his farm, if he can get a tenant that will farm well. Though he well knows that he could make more by hiring his work done, yet he wishes to relieve himself and family from the trouble of taking care of the farm, and hired help. Now why can't a tenant, if a young man with a small family, take that farm and go on and make money, and, at the same time, keep the land in good condition? The owner made money, and kept the land improving; why may not the tenant make money, and at least keep the land in its present condition? I see no reason why he can't, nor do I believe there is any—except poor management.

"Again, let those that think they can't afford to farm well on another man's land look to England. Much of the best farming in that, or perhaps any other country, is tenant farming. Not only does the tenant have to pay enough, in rent and taxes, to buy land in many sections of this country, but he spends thousands in manuring, and other improvements. Indeed, it is said that his rent, taxes, and other farm expenses are so large, that he is obliged to cultivate his land in the best manner; that he could not get along without doing so. Yet he lives well and makes money; and it is said that many tenant farmers do so well, and are so well satisfied, that they prefer remaining tenant farmers, even after having made money enough to buy and have land of their own.

"Now allow me to ask, why may not something like this be the case here? Why may not an American be a good tenant farmer as well as an Englishman? Have not our young men as much enterprise, intelligence, and ability, as the same class anywhere, and are they not as anxious to make money and go ahead? Then why not make the business of taking or renting land, one of the best courses a young man can take to get a start in the world; instead, as is too often the case, making it a losing business for all concerned?

"But it will be said, taking for granted that tenant farming may be made to pay well, how are all that may wish to, going to get land to work? It has already been intimated that the demand for farms was much greater than the supply; hence it must be admitted that, though it may be a very good way to get a farm, for those lucky enough to get a good farm to work, yet many will fail, because there are not farms enough to be let to supply the demand. This being the case, we will pass on to consider the third and last course proposed for a young man to pursue in order to get a farm.

"This course, as well as taking land, is more particularly calculated for a married man. Though the single man that is able to get good wages and steady employment, may do very well, yet when he gets married he wants a home, and generally the sooner he gets one of his own, the better. Time and space forbid giving even a tithe of the reasons why every man should have a home of his own. All are more or less familiar with these reasons; and as undoubtedly one of the principal reasons why G. B. S. wishes to get a farm, is to have a home, I need only state that I have found, both by experience and observation, that a small piece of good land, even though there may be but a few acres, is a great help to a laboring man that wishes to get along in the world. Here, again, time will not admit of referring to the many instances of large amounts of produce grown on a few acres of land, that I have come across in reading and observation. But I must pass on, only stating that few young men are aware of how small a place may be made to give them more net profit for their labor than they can realize by working out.

"Another advantage in having a small place is, that it will enable the owner to do something at both farming on his own land and working out. Whenever I have seen this step in advance (which it surely is) taken by a man of industry and economy, I have always observed that he went ahead much faster than he did when depending on his labor alone. So, too, a small farmer will often find chances to take land by the piece, of farmers having more than they can or wish to cultivate, thus enabling him to add to his farming operations sufficiently to give him all the business he can attend to, and giving him quite respectable profits.

"But leaving working out, or taking land, out of the question, few that have not tried it, or investigated the subject, are aware of how few acres will keep a man profitably employed during the busy season. In a former article on farming on a small farm, I have given estimates of what can be raised on ten acres, and also on twenty acres. These estimates, though much less than is often realized, will give a good idea of what may be raised on a small place, it being kept in mind that by changing works for team-work, the owner may do nearly all of the work himself, making his expenses out scarcely any thing to speak of, and enabling him to realize the full benefit of all he raises.

"Another great advantage in getting a piece of land as soon as possible is, it forms a beginning—a something to add to. Young men when working out, or working land, and having no particular present need or use for their money, are apt to spend it, or allow it to slip away for something that, in their circumstances, they might better do without. But if they have land that they wish to finish paying for, or to make improvements on, or, finding their little place insufficient for their wants, and aided by the stimulating effects of actual ownership, they are anxious for more land, they will be sure to save all they can to buy more with. When this course is once fairly entered on, they will be pretty sure to follow it up until they are each one the owner of a good farm.

"Before concluding, I wish to present one or two considerations that are very important for young men of limited means that wish to get farms. The first is, that in taking any course that will be open to them, they may not be able to make money as fast at the beginning as may be deemed desirable. It is very natural for young men to make large calculations at the start. They have a very laudable ambition to go ahead and make something, and be somebody; hence they are apt to think that any course they may be able to take is too slow to ever accomplish any thing. But this is a mistaken idea. Let any one that doubts this sit down and reckon up what a man that earns $100 over a living every year from the time he is 21 until he is 50, and puts it at interest at 7 per cent., adding the interest to the principal each year, will have when he is 50 years old—or say in 30 years. I say, let him do this and he will be surprised to learn that he may be a comparatively rich man, by taking this course, when he is 50 years of age. As a further illustration of this fact, I will mention a few instances that have come under my own observation, one of a man that died worth over $10,000 in cash, that made it, all but a small legacy, by working out and the interest on his money. Another, that is now some 35 or 36 years old, that has between $3,000 and $4,000, all made by working out, and the accruing interest on his wages. And yet another that saved $900 in six years. All this shows most conclusively that, though either of the courses I have pointed out may seem rather a slow way on the start, yet, if persevered in, and all of the money, as fast as realized, invested in some manner whereby the interest is sure to be realized, they are sure to lead to the desired success,—while hundreds, perhaps thousands, have done a great deal better than this by investing their labor and money in farming in such a manner as to realize much larger profits.

"The other consideration, with which I shall conclude, is

that every young man that wishes to succeed should make himself familiar with the agricultural literature of the day. He should not only read and keep for future reference some of the best agricultural journals of the day (of which I wish to say that the *Country Gentleman* stands at the head), but he should be familiar with some of the best practical works on farming in the country. He will find this a great advantage, if he works out, in enabling him not only to work to much greater profit and advantage to his employer, and thus getting the extra wages that will be his due for highly intelligent labor, but in showing him how the knowledge gained by his present experience may be turned to his future benefit when farming for himself. Or, if taking or renting a farm, in learning how to manage it to the best advantage, both as regards present and future profits. Or, if farming on a small place, not only in learning what may be and has been done on a little farm like his own, in different sections of the country, but in learning how he may manage his few acres to the best possible advantage. But above all else, he will find it of the greatest advantage in enkindling in his mind an ardor for, and an enthusiasm in the business of farming, that, enabling him to triumph over every obstacle, will be sure, sooner or later, to bring him to the desired haven of success."

These original suggestions drew forth a second reply in the same paper, under the signature of "R. S. F.," as follows:

"A correspondent inquires, *How he can get a farm without money or capital to buy it with or to conduct the business of farming?* I can answer his question in two words. *Take mine;* with this proviso, however, that he understands practically and thoroughly the profitable management of a farm, and has a character in all respects equal to his practice and his knowledge. He is looking for a farm; I am looking for a farmer who can take hold of the soil in a way to improve it and his own condition at the same time. My farm lies vacant and unimproved, because no one appears that can satisfy me of his capacity to do this. I can find plenty of men who would be glad to buy the farm upon a credit, but who would never pay for it, and who would tease, depreciate, and worry it to no purpose. I can find others who are ready to rent it for a stipulated sum, or upon shares; but no one has ever appeared that possessed sufficient qualifications to manage the business, to keep the farm improving, and to do this. If he prospered, it would be tolerably certain that his success was *at the expense* and not *by aid* of the land.

"This farm is accessible to good markets, and contains six hundred acres of land of every variety, clay and light; plenty of meadow, salt and fresh; abundantly supplied with wood; with never-failing sources of water, and surrounded with schools, churches, etc. Now I am willing to sell this farm upon a reasonable credit; or, I am willing to let it to a responsible, improving tenant, at a low price, whenever I can find a man of the right sort to take it, with a condition attached to the lease, that the tenant shall have the right to purchase it at an agreed price within a given period.

"A tenant usually grows rich on a farm, for the reason that he usually goes upon it with a view of laying up enough money to pay for a farm of his own elsewhere. He carries from the hired farm in his breeches pocket all the *scrapeable* value he can; it has been made poor to make him rich—in other words, he has transferred the fertility of the one to the fields of the other by a sort of electrotyping process, whose transmutation in soils, are in such hands as sure as they are by the labors of chemistry in metals.

"Now, Messrs. Editors, when you have followed me thus far, should I stop, you would editorially ask, Why not advertise your wants in the *Country Gentleman?* I answer, because if I did, I should have no end of applications from this very class of people that I wish to save my land from. I am diligently seeking for an experienced, money-making, land-improving farmer; but, in the mean time, my house, now old and shabby, is rotting away, and my barns will soon follow in one general decaying, destructive sweep. Shall we ever have schools of agriculture, from whose portals, as they graduate, one can find a competent agent, tenant, or the purchaser of a farm, who has learned the art of making it pay for itself?"

CHAPTER IV.

More Opinions and Experiences—Some Objections—Additional Light—Encouraging the Young—A personal History—Getting an Illinois Farm—One Example—Good Suggestions—Buying and going in Debt—Value of the Discussion.

The discussion thus opened drew out, as may be supposed, the views of other practical men to elucidate the important question as to the best way of getting a farm. The following is the commentary of another intelligent observer, Mr. J. W. Colburn, of Springfield, Vermont. Referring to the suggestions made by "F.," as quoted in the preceding chapter, he says:

"His advice cannot but be regarded, by those to whom it was intended to benefit, as very sensible, and in the main correct. He points out three ways to be pursued to accomplish the object sought for, viz.: Working out for wages, taking farms upon shares, and beginning with a few acres at first, enlarging as means are saved to invest, seeming rather to give the preference to this last method over the two first. Circumstances, with regard to land and labor, may be such in his locality as to make his views correct; but with all due deference to his opinion, to suit the locality in which I reside, I should ask him to reverse his opinion, and put the working out for wages to get a start in life at the head of his three ways to get a farm, as decidedly preferable to either of the other two.

"If the first thing that a young man thinks of and must

do, after arriving to the years of his majority, and destitute of means, is to get married, as is often the case, then he must do the next best way that he can—take a farm on shares, or purchase a few acres; but it will be but a few acres, in four cases out of five, that he will ever be likely to pay for and own. I know it is a divine command to 'multiply and replenish the earth,' and that it is good for a man to provide himself with a 'help-meet,' but we are told also that there are times and seasons for all things, which may be interpreted to mean a proper time and season, leaving every man to be his own judge and monitor as to what is best and proper to suit his own case and circumstances.

"But how few there are that can look back and review the past events of their lives, the most prominent ones, such as have influenced their career for good or evil, and say they have acted wisely or judged correctly! Man is the creature of impulse, more or less improvident and reckless, acting from influences that surround him, without stopping to take the 'sober second thought,' and often makes shipwreck of his future well-being in life, when by a more prudent, discreet, and judicious course, all might have been smooth, bright, and prosperous. It is desirable and pleasant for a young man to seek a home—to make a home of his own. The associations attached to that word, *home*, sink deep in the human heart. But no young man should be in such haste to realize this inestimable blessing, as to turn it into a curse. A home that is surrounded with poverty and want is no home; and more particularly is it so when not brought about by any unavoidable misfortune, but the result of miscalculation, acting from other motives than those of a prudent and discreet foresight. I never can advise any young man to incumber himself with a young family to support until he has the means, and can see his way clear to do it creditably and comfortably.

"F. gives us an instance of a man that died 'worth $10,000 in cash, that made it all but a small legacy by working out and the interest on his money.' 'Another that is 36 years old, that has between $3,000 and $4,000, all made by working and the accruing interest on his wages.' 'And yet another that saved $900 in six years.' These cases show most conclusively that working out on a farm by a young man starting without means, who is determined to own a farm, is the shortest and surest way to accomplish that praiseworthy and desirable object. Had either, or all of these cases, detested working out, as many young men do, and got married at the start, and relied upon taking farms upon shares, or upon running in debt for a few acres, think you at the same period in after-life they could have shown these results? I tell you they could not, but probably would have seen an old age of destitution and want. I know at the present day there is an antipathy among our young men to working out on a farm. As F. says, it is useless to talk to them about working out ten or fifteen years in order to enable themselves to become owners of good farms. The process is too slow; some more rapid way must be contrived; but the effort rarely proves successful, and the farm is very seldom owned. In olden times, farms, and good ones too, were often obtained by this patient and persistent industry, coupled with strict and rigid economy. True, they cost less then than now, but farm-wages comparatively were as much lower then than now as the farms were lower, while every article of clothing was 50 per cent. higher than they were previous to our present war prices.

"I have known many independent farmers that commenced life by working out for wages, but a precious few that commenced by taking land upon shares, or by purchasing a few acres, that ever were the owners of productive farms. This working land upon shares, or the product of a few acres, must necessarily be absorbed by the support of a family, while the young farm-laborer should, and generally does, remain without a family until he is able to purchase a farm and see his way clear to pay for it. While working out, it is not convenient to have a family, but almost indispensable when working a farm upon shares. F. says that when a farm is to be let on shares in his section, six or eight applicants appear, to improve the opportunity; and why is it so? The reason is obvious. Having commenced in this way, they do not rise above it. They can support their families, and that is all they can do, while the single man, at work for wages, is placing his hundreds at interest.

"It is not in good taste for a man to say much about himself, but if your readers will pardon the egotism, I will relate, briefly as possible, a sketch of personal experience. The writer remembers well, when a young lad, of listening to the stories of several farmers, most of whom he was in the habit of drawing out by questioning and inquiries, as to how they commenced life, and how they obtained so much property, etc. A rebuff from no one of them for being an impertinent boy was ever experienced, believing now that they deemed his object to be something more and better than an idle curiosity. The information obtained in this way served as an exciting purpose with the young inquirer to stimulate his ambition to do likewise, and become, like them, a man among men of independent means. He pursued the same course that he now recommends to others. Working out upon the farm was the beginning; the most untiring application and persistent industry for thirty years, has brought results beyond his most sanguine expectations, and he has no cause to regret his early determined resolution.

"Why is this farm-work for hire so much to be dreaded? It is frequently the case that the hired man, though in one sense a farm-servant, works no harder than his employer; he does not have the care and perplexity on his mind that the owner does; fares equally with the family, and at the expiration of the year has more surplus cash to put at interest than the man who has employed him. What has been done by some men can be done over again by others. If our fathers commenced in this small and patient way, and became men of wealth, of character, and of influence and standing in society, so can our young men of the present day, and in many less years than it took them to gain their position.

"The reader of modern history will recollect that when the first Napoleon appeared at the head of the French army in Italy, then but 27 years old, that he astonished the old veteran generals of Europe by his new military tactics. His rapid movements, his furious onsets, they were not prepared to expect, and much less to resist. Where did this prodigy learn this new art of war? When in captivity on the desolate rock of St. Helena he let out the secret. He had studied closely the history of Alexander the Great. It is said that he always carried about with him the life of this great Grecian conqueror. He knew that the world as it was in that ancient day, never could have been conquered only by the rapid and determined means brought to bear on the enemies of Greece; and he reasoned with himself, and correctly, too, that what man had done, man could do again; and firmly relying on this principle, he put the same means in force, varying as the means of warfare had varied, but keeping steadily in view the rapid blows and the determined energy and zeal that gave the Grecian universal empire. To a very great extent he did over again the work of Alexander, and caused every throne in Europe to tremble, though all combined, finally overthrew him.

"This is a far-fetched analogy, but it is none the less true.

What man has done, man can do again. It only requires the determined will, the energetic hand, the unremtted perseverance, and the patient, long-enduring application, and almost any object, legitimate and honorable in its end, can be reached. I say then to young men, beginning life without means, having a taste for the business and desirous of owning farms, 'you can do it if you will.' You cannot be any more destitute at the start than was he who now addresses you. The times and circumstances now are vastly in your favor, over those of forty years ago, and I bid you God-speed."

This expression of opinion was followed by a rejoinder from "F.," to this effect:

"MESSRS. EDITORS—In former volumes of the *Country Gentleman* there have been several articles on buying farms, which were mainly calculated to benefit those with plenty of money to buy with; while there has been very little, if any thing, given, calculated to show those with limited means how they can buy to the best advantage. So, too, those writing on this subject have generally seemed to consider or take it for granted, that it will not do to run in debt for land, although it is a very common practice in most, if not all parts of the country, to do so.

"Now, without stopping to consider whether it will do to run in debt for a farm or not, or the amount of capital a man should have to commence with on a given number of acres, I shall take it for granted that as a great many have run in debt more or less for farms, which they have not only succeeded in paying for from the produce of the land, but have also, by the same means, attained to forehanded and often independent circumstances; and as the example set by such men is constantly before those anxious to attain to the same station in life, a great many will still continue to run in debt for land, any thing I might be able to say to the contrary notwithstanding.

"True, many fail, and not a few farms are badly run down and injured by men in debt; yet this need not necessarily be the case, but may rather be considered the result of bad management in buying, cultivation, etc. But, however this may be, it will not prevent others from running in debt. So, that instead of a fruitless endeavor to persuade men to not run in debt, I believe it is better to try to assist those that may find it best to do so.

"Of course no specific rules, but only general directions, can be given as to when a man's pecuniary means will admit of his buying a farm. There is so much difference in the price of land in different sections, as well as in its productiveness, that while in some new sections where land is cheap and a long time given to pay in, a man with a few hundred dollars may buy one hundred acres or more, in many of the older sections he would need as many thousands to buy the same amount of land. But leaving new lands out of the question, when should a man buy in those sections where land ranges from $20 to $100 or more an acre? As a general rule, I believe he may buy when he is able to pay half down for a farm, and stock it with a moderate allowance of farm-stock suitable to the system of farming he intends to pursue, together with suitable teams, tools, etc., to begin with. And as a man that intends to pay a heavy land debt should not allow small debts to accumulate on his hands, so he should never begin by making a heavy debt on his land, and another large one for teams, stock, etc.

"Again, the amount it will do to run in debt must depend, in a great measure, on the time and chance to pay, the buyer may be able to get. As when $2,000 or $3,000 are wanted in three or four years on 100 acres, it would be well to hesitate, while it might be safe to agree to pay that amount in ten years; so, too, it will be found more difficult to pay a given amount in large payments of from $500 to $1,000 each, than it will to pay the same amount in $100 or $200 payments. The best way, where a large debt has to be made, is to take a deed and give a mortgage, payable annually in such sums as the buyer is sure he can pay besides interest, with the privilege of making the payments faster, should he wish to do so. Then, by managing to keep one or two payments ahead, he will never be distressed to make his payments in bad seasons, nor be obliged to sell when produce is ruinously low. This is substantially the course taken in buying my present farm, and I have found the advantages named, as well as others that might be given, a great help in making my payments.

"Another point that should be well considered, is what kind of a farm is it best to buy? It is generally said, that it is best to buy a good farm, in a good state of cultivation, with good buildings, etc. This is undoubtedly true as regards those with plenty of money to buy with; and such farms are sometimes the cheapest for those with limited means. There are many farms where the land—naturally good—is in a fair to good state of cultivation, with moderate but good comfortable buildings, that can be bought for reasonable prices. Such farms, though they may not be cried up as the best, it is always safe to buy. While those farms that have the reputation of being the very best in the country, and are cried up to the highest rate, it may be well to pause before purchasing. It may be well to consider how much of this credit of being an extra good farm may be due to an extra good farmer, and also whether the high price such a farm is sure to be held at, is wholly due to the superior condition of the soil, buildings, etc., or whether a part of it may not be ascribed to the high reputation in which the farm is held.

"But there is another class of farms that I wish to more particularly recommend to those wishing to make their money go as far as possible in buying land—which is those farms that, though naturally good, have been run down more or less. There are many of this class of farms in all parts of the country. They have been and are usually occupied by poor farmers, that—whether owners or tenants—generally do the work to halves, and only raise half crops, and as they seed down but little, if any, their farms soon have a very poor, barren appearance, causing their reputation to go down very fast, and often causing them to be sold for much less than their real value. That is, while the high reputation of extra farms causes them to sell for all or perhaps more than they are really worth, the bad name and poor appearance of badly run farms, often leads to their being sold for much less than their actual value. True, the farmer that commences on a badly run farm, will have to adopt some course of improvement by which the soil may be again made productive, or it would not do to run in debt for the farm. But it will not be very difficult to do this if the land is naturally good, and has only been run down by poor cultivation and neglect. I hold it an undoubted fact, that naturally good land cannot be thoroughly run down and worn out without a more thorough working and course of cultivation than such farms usually receive; and also, that much of the credit that is often ascribed to this or that course of improvement, is due to the latent goodness of the soil, developed by more thorough cultivation. Of course, before buying, we should be sure that the land is naturally good, and that its present bad appearance is due to poor cultivation and neglect, instead of a more thorough course of cropping on the skinning system, by which the soil is actually worn out. Yet such cases as the last are rare, as really thorough cultivation is generally found in connection with some system or course of management by which the soil is improved instead of being run down.

"There are other reasons for buying this class of farms by men of limited means. One is, that such farms not being generally as salable as those in good condition, a much longer and better chance to pay may usually be obtained. Another is, that by good management and cultivation, such farm may soon be made to bear an altogether different appearance, which, with the good crops that will be raised, will be sure to greatly enhance the character and reputation of the farm, and make it sell for a handsome advance on the cost; while its enhanced value and productiveness would be no less real and satisfactory should the owner not wish to sell. But some one will say, if badly run farms can be improved and made to produce good crops, by men more or less in debt, what thousands will wish to know is, how it can be done? This I shall endeavor to show in another communication."

This communication drew the following from a hitherto silent observer of the discussion, Mr. Jonathan Talcott, of Rome, N. Y. After stating that he has carefully read the articles of Mr. Colburn and of "F.," he says:

"I beg leave to dissent from their views as expressed in their communications, and take the middle ground, as spoken of in the articles referred to, and shall advocate it, and advise such young men as Mr. Colburn speaks of to adopt that plan in getting a farm.

"Let us briefly look at the qualifications mentioned. He is to be ambitious and energetic. Now, we suppose Mr.

Colburn means a laudable ambition, such as is becoming any young man to possess, and energy also to persevere under difficulty (if need be); and, in his closing sentence, he adds, 'What man has done, man may do again.' Also, 'that it requires the determined will, the energetic hand, the unremitted perseverance, and the patient, long-enduring application, and almost any object legitimate and honorable in its end can be reached.'

"Such, then, are the qualifications, if I rightly understand Mr. Colburn, that the young man wishing a farm must possess, and such Mr. Colburn advises—in order to become the owner of a farm in the shortest possible time—to work out by the month on the farm for a period of from ten to fifteen years, as circumstances may require; and to substantiate his position, he quotes the following sentence from F.'s communication in the *Country Gentleman:* 'F. gives us an instance of a man that died worth $10,000 in cash, who made it all but a small legacy by working out, and the interest on his money. Another that is 36 years old, that has between $3,000 and $4,000, all made by working, and the accruing interest on his wages. And yet another that saved $900 in six years.' Mr. Colburn then says: 'These cases show most conclusively, that working out on a farm by a young man starting without means, who is determined to own a farm, is the shortest and surest way to accomplish that praiseworthy and desirable object.' Also, he adds: 'Had either or all of these cases detested working out as many young men do, and got married at the start, and relied upon taking farms upon shares, or upon running in debt for a few acres, think you at the same period in after life they could have shown these results?' Mr. Colburn then answers the question decidedly in the negative, saying—'I tell you they could not, but probably would have seen an old age of destitution and want.' The language used by Mr. Colburn is very strong and decided, and perhaps it may be considered presumption in me to gainsay it. Viewing it in the light of my own experience and observation, I consider his views, as expressed in the answer quoted, as containing errors, and calculated to mislead those whom he desires to benefit by the advice given in his letter from which I have taken the extracts already quoted; and relying on Mr. Colburn's forbearance towards one who may differ from him in opinion, will briefly give my views on the subject.

"F. does not say that the men mentioned made the sums credited to them by working out on a farm as farm hands, although Mr. Colburn assumes as much, and the price mentioned by F. in his first letter, when he quotes from Mr. Colburn's previous writing, at $15 per month for the year, is higher than the average in this county for the past twenty years. I think $150 per year is full an average for the best hands for the time mentioned; some of the time above, and some below that price, some not getting that price; but we are talking of the ambitious and energetic young man who is determined to become the owner of a farm in the shortest possible time, who has it to earn by his own labor.

"We see, from what has been written by F., and also by Mr. Colburn, that from $100 to $150 per year can be saved by such men under the most favorable circumstances; probably $120 would be more than an average in the cases that have fallen under their observation, and in this vicinity it would be under that estimate.

"Now we will look at the other side of the picture, keeping in view in the mean time the qualifications that the young man must possess. We will suppose him to be 21 years of age, with good health and all the requisites mentioned by Mr. Colburn, and that he has formed associations with some young lady of his acquaintance, who also possesses the same qualifications, but she also is without means to purchase a farm, although that is the object of their ambition, to attain which they are ready and willing to unite their interests in the realities of wedded life, and commence in earnest to acquire the desired object. In the light that Mr. Colburn views it, they will probably fail; but viewing it in the position that I take, they will succeed, and much sooner than they would if both worked for wages, having the same end in view, viz., a good farm.

"Such a couple can easily secure a good farm on shares from a good landlord in this State, and probably in most of the States of the Union, and instead of saving $100 to $150 per year, can, with the blessing of Providence, lay by from $300 to $500 per year, with more advantage on their side than to remain single. For instance, if they remain single, and are sick, all income is stopped, while if either is sick when married, and on a farm, crops and animals are still growing, and there is one (if only one is sick) to look to the interests of both. Besides, how much more congenial to human nature to have a companion to share with each other our joys and sorrows, either in prosperity or adversity! Also, what young lady of the qualifications mentioned would not prefer to join her destiny with a young man of like qualifications at once, and begin the great battle of life in earnest, and having conquered, as they surely will, all obstaclés at the age of thirty or forty years, look back with pleasure upon their commencement and struggle for a farm, and feel that, with their united efforts, with the blessing of a kind and heavenly Father resting upon them, they have succeeded beyond their most sanguine expectations. Perhaps, too, at that age sons and daughters may have grown up around their fireside to gladden their hearts, who will soon go forth in the wide world to conquer the obstacles, either real or imaginary, that may arise in their path. Think you there would have been as much enjoyment and genuine home-feeling had they remained single in the mean time? I think not. I have in my memory at this time numerous instances, such as I have mentioned, when, at less than fifty years of age, persons that were married in early life, and have worked farms upon shares, have made their thousands, and purchased farms of their own, and are still in the prime of life; also some that have told the writer of this that they did not want to own a farm, as they could make more money in working land belonging to others than to be the owners themselves.

"Such have been my observations while engaged in working out the problem for myself; for, at eighteen years of age my prospects, in regard to property, were dark enough (my father having died ten years previous without property; consequently, myself and mother were left to rely upon our own exertions for a livelihood). For three years I worked out on a farm for wages; that is, from the time I was eighteen years old till I was twenty-one. I will not say that I had the qualifications mentioned by Mr. Colburn; but I may say that I saved what I earned, and tried to do my best for my employer. At the age of twenty-one years I made arrangements to take a farm with a view of purchasing. Soon after, I married a young lady that knew my circumstances, and that the farm had to be earned if we ever wanted one. We commenced in earnest. Before we were thirty years of age the desired object was attained, and we were better off, in dollars and cents, than the cases cited by F., which Mr. Colburn quotes; besides, we had been at home all the time while we were engaged in paying for it.

"As for the expenses absorbing all the income in working

a farm upon shares, as mentioned by Mr. Colburn, that has not been the case with those of my acquaintance. The expenses in such cases with such a couple will be light, and they will save from $3 to $5 in such a situation, where but $1 would have been saved by the working-out system.

"Now, I do not deny that many young persons will fail to accomplish the results I have stated; but such would fail if single, not having the required qualifications. I could mention cases where, when single, not a dollar was saved, when as soon as the same young men were married their earnings were saved, and a competence was the result in a short time.

"The expenses of a family are never so light as at the commencement, and any young woman with the qualifications to match the young man, as given by Mr. Colburn, will, instead of adding to his burdens, lighten them in very many ways, which I need not mention at this time; besides, he will doubtless have the blessings bestowed on him, designed by his Creator, when he said, 'it is not good for man to be alone,' who also created 'an help-meet for him.' There is no doubt but the same injunction is now resting on the human family as in the earliest stages of the world, and we shall do no wrong by heeding the injunction.

"In conclusion, I would advise all such young men as mentioned by Mr. Colburn, and all others that think they have the qualifications mentioned, that the sooner they get married after they arrive at the age of twenty-one years, and a suitable situation offers to take a desirable farm, the sooner they will become the owners of such a farm for themselves; and that, instead of a young woman being an incumbrance, as Mr. Colburn intimates, she will be a real help, and they will save from their united earnings in the ratio I have mentioned; besides, she will be a sharer in his joys, and grief, if he has any, which he doubtless will, as trials are the common lot of humanity. She will also cheer him on in all his successes, looking at the object to be attained on its brightest side, thus proving most conclusively her ability to perform her part in the task they have allotted themselves—which, with God's blessing, they will surely accomplish."

A writer, under the signature of "A Farmer's Son," now threw into the common stock of information the following brief summary of a very interesting personal history:

"Having read in the *Country Gentleman* several ways for a young man, desirous of obtaining a livelihood by farming, to do, I thought perhaps a few ideas I might suggest would not be out of the way. Although young and inexperienced myself, in the ways of working and by the means of which a farm is obtained, I have often heard my father speak of his experience, some of which I will briefly relate. At fifteen years his mind was fully made up to be a farmer. To that he devoted his energies, and boy though he was, was fully assured that he would have no other vocation. At eighteen he bid adieu to father and mother, and started with nothing but an axe, which was all the kind parent could give, but his blessing, and a piece of bread and cheese from the thoughtful mother. He left the parental homestead, travelled thirty miles, there found employment, and from that day to this never has known want. For the next five years he labored partly by the month, and also by working farms on shares. In those days, when working a farm on shares, you boarded with the family, including washing, and had one-third of the profit. In the next five years he laid up $500—was then married, bought a small farm for $750, paid $250 down, with five years to pay the balance. He worked it eight years, then sold, and was worth at that time $2,100. Worked a farm on shares for two years—was then worth $3,100. Then bought a farm for $4,500, having it so arranged that the payments would be made from the grain and meat raised on the farm. When that was paid for, he sold again and bought another for $8,200. By improving in fencing and building, the farm is now worth $13,000. Many young men, who commenced with nothing, have now good homes, surrounded with all the comforts of life. Working a farm on shares, he thinks is quite as profitable for a young man as working by the month."

The importance of avoiding an unfavorable location, will be seen by attending to the description of farm-lands in the neighborhood of Jamestown, New York, as given by Mr. W. H. Benson, in the following article:

"Again, in regard to the best way for a poor young man to get a farm: I think the discussion on this subject has not yet touched bottom, and that much more might be said that would be of advantage, at least to some *lone* one. Working out by the month, and earning enough to buy a farm, would, in my humble opinion, bring a *young* man so far past that title, that, in nine cases out of ten, he would go down to the grave 'gray and sorrowing,' with the object of his lifetime still far in the distance. As to taking farms or working land on shares, I think that would be the best and only way: but even then it must be sought under favorable circumstances. In this locality it would be uphill business, and the reason why is this: In the first place, the price of farming land is about double its real value for farming purposes. Secondly, the landowner asks double the rent that the tenant can afford to pay, and if the tenant takes a farm on shares, the bargain will be something like this: The farm will be poor, for no good ones are to be let. The young man must do all the work, pay all the taxes, repair miles of old tumble-down fences, and make a certain amount of new—repair and fix up the old house and barns, build his own sheds, hog-pens, &c.; cut, in the aggregate, about ten acres of brush around stumps, in fence corners, &c., then put in all his *spare* time to picking up stone and rubbish, and drawing them from the meadows.

"Then he must find his own team, tools, and stock (if he has any), keep it on his own half of the hay and grain, and pasture for the owner of the farm an equal amount of stock. He must then give one-half of all the produce of the farm, and will be allowed to stay but one year; and why? because the farm is now in shape, and the owner can *afford* to work it himself, or rent it to some one that will agree to pay twice the rent he will ever be able to. Would it not be a good way for a young man to rent a good farm, of a good man, in a good locality (where he can make it an object for himself and landlord also), with the privilege of buying, and the rent, in case of purchase, to apply on the purchase-money? At least that is the opinion of *one* who intends to try it when a favorable opportunity occurs."

So far the West had taken no part in this discussion, though evidently watching its progress, but now she claims a hearing, as will appear from the following remarks of Mr. J. B. Porterfield, of Champaign, Ill. His representations are in striking contrast with the discouragements suggested by Mr. Benson:

"We have been some amused at the correspondence which has been published in your paper since the inquiry was made as to the best way for a poor man to get a farm.

Mr. Benson, of Jamestown, N. Y., gives a gloomy account indeed of those who intend to *try* to buy a farm. The idea of giving one-half of all raised, besides hauling rocks, cutting so many broad acres of brush, building pig-pens, making sheds, &c., all for the privilege of living on a poor farm—(Mr. B. assures us that none but poor ones are offered foi rent)—looks more like a joke than a reality to a Western man.

"I am not accustomed to writing letters for publication, nor am I in the habit of giving advice, but after reading such gloomy accounts as those given since the inquiry was first made, I determined to try my hand at it. And now I would say to all those who have the pluck to try to get a farm under such disadvantageous circumstances as Mr. Benson speaks of, to pull up stakes and come to Illinois, where we will bid him a hearty welcome, where there are no stones to pick, nor brush to cut around stumps and in fence corners, and where, if you persist in renting, your landlord will ask one-third of the grain raised, pay all taxes himself, and furnish you with house, fuel, &c. But still a better way would be to buy a farm for yourself, should it be only 40 acres. Mind, our lands are not encumbered with rock, brush, &c. But your whole 40 acres is a perfect garden spot, and all the stock you may have can roam on the prairies at large, which are covered with a luxuriant growth of grass of the best kind for either pasture or hay.

"Now, you will probably be ready to inquire what such land can be bought for? It will cost from $3 to $10 per acre, unimproved, and from $5 to $20 improved, and this on almost any time the purchaser may desire. The Illinois Central Railroad Company have, as you are aware, obtained from the general Government a large grant of land to aid them in the construction of their road, which has been built, and hence the title is perfect. The company are selling their lands now, and thousands of just such men as your correspondent seems to be are availing themselves of the liberal terms to secure to themselves homes, and thus avoid renting lands, or in other words, becoming landlords instead of being tenants. Their terms are as follows:

"Say Mr. B. buys a farm of 40 acres, at $5 per acre—$200. If paid in cash down, a reduction of 20 per cent. will be made—$160, which will be the entire cost. If the purchaser choose to avail himself of the time given, it would be thus: $200 at 6 per cent. for four years, $12 per year; at the end of said time one quarter of the principal comes due—$50, which amount comes due annually until the principal is paid;—thus enabling Mr. B. to make the price of his new home from the land itself. According to an act of our State Legislature, all their lands are exempt from taxation until paid in full, which is of itself considerable inducement at a time like the present, when high taxes stare every man in the face.

"I have thus endeavored to give the outline, hoping it may lead those who are laboring so hard to become lords of the soil, to examine into the matter, at least. I would say further: This is a healthy place; good pure water is abundant; society is good. Now, gentlemen, let me advise you to quit renting lands in New York, or any other place where you have such hard taskmasters, and where the lords of the soil ask double what the lands are worth. Should you happen to ever be able to buy yourselves homes, come West; we invite you to no mean country, but to the garden of the world. This is admitted on all hands. Illinois is but an infant in years; still she is the fourth State in population, and in the amount of grain raised, and the amount of cattle and hogs, the first in the family of States. We will welcome you with open arms to our broad prairies and fertile soil, and where the inhabitants are loyal to the best form of government the sun ever shone upon, where the people respond to their chief executive whenever he calls; and I will here say, and I must confess with feelings of pride, which the past proves, our people are ready and willing to leave friends and pleasant homes, and rally at our country's call. It matters little whether our quota be 30,000, 40,000, 50,000, or 100,000 men, we are ready, and will respond. Our cattle are all fat, our hogs also; our granaries are full of small grain, which now command a good price. Our corn is all worked, and nothing now remains but to take it off. Our mothers, wives, sisters, and daughters say, let all the men go if necessary; we will gather the corn. Thus you see we are ready."

A cheering example of success, with continued comfort during an entire life, is contributed to the discussion by Mr. N. Reed, of Amenia, New York:

"Those examples of successful farming where young men have been able, in a few years, to pay for and improve their farms, seem, to many, extreme cases of success, and exceptions to the general rule. The more usual conditions of attaining the possession of a good farm, are many years of industry and careful management. The profits of farming are moderate, and the acquisition of a good estate by this calling requires more patience than by any other. And this is the reason why many young men turn from this to some more promising profession. They are in haste to possess the means of setting up a comfortable establishment, and cannot think of waiting for years for an independence, and turn therefore to some business which promises a fortune in what they are pleased to call a reasonable time. They mean to have a farm and gratify their rural tastes, when they shall have made money enough by some profitable business to be able to do so.

"It is not my purpose now to show how often these young men are disappointed, and how they are deceived in their views of the enjoyment of rural life. My object is to give a word of encouragement to those who may be tempted to shun the slow way of getting an estate, and my lesson shall be drawn from the experience and observation of more than twenty-five years. The experience of a large number within my observation, during this time, may be considered embodied in one case, which is probably a fair sample of a very large number all over the land.

"A young man began his career by working for his father at stipulated wages, which he continued a few years, until his ability was fully tested, when he took the management of a farm on shares. The time at length came when he would have a farm of his own, which he purchased with much solicitude and many fears of ultimate success in paying for it. His former accumulations, together with a small patrimony from his father's estate, was not sufficient to pay for half the cost of the farm and stock, and he often wished himself on a smaller place, and out of debt. The farm he purchased, like many others of that day, had been pretty hardly run with crops of grain, and was much out of repair. There was an enterprise requiring good courage and perseverance. Our farmer had three principal things to do: he had his family to support, which was not small, his debt to pay, and his farm to improve. It is sufficient to say that he carried these on together every year. There was no year but at the end of it he had, besides paying the interest and improving his farm, paid something upon the principal, generally however a little less than he had anticipated. He had not only these things to do, but he must sustain the social position which his family and his education entitled him to. His civil, social, and religious relations must be maintained. Not one of these was neglected; he stood in his place as an American citizen, and took upon himself all the duties and responsibilities of his station. In this respect he avoided an error which most young men in debt fall into, who in their impatience are ready to forego almost the comforts of life, ignore most important social relations, and leave all improvements till they are out of debt. They are willing to deny themselves and their families all the elegancies of life, and make themselves mere drudges to obtain first a competence; and, when this is accomplished, they find themselves unfitted by their habits and associations for a true enjoyment of what they possess, have become sordid, and are only satisfied with increase of gain. But our farmer improved his place with many tasteful though simple embellishments, and his mind by reading and good society, and this without any great expense of time and money. He rejected the principle that a man ought to make as much money as he can.

"He accomplished what he undertook, in paying for his farm. Not quite as soon as he expected (the gray hairs begin to crown his head), for he had his share of the reverses of business, and it might seem a long time; but when he had paid his debt, he had a complete farm, a good estate, a competence. For, as I said, he improved his farm yearly, so that the productive capacity of it is more than doubled, and the net profits are in still greater proportion; the fences are good, and the buildings greatly enlarged and improved, and the stock of the farm increased in number and value. So that what cost him ten thousand dollars is now worth more than twenty thousand dollars.

"Now, the very thing which our fast young men picture to themselves as the desired end of all their anxious toil and hazardous speculation, a quiet enjoyment of rural life, he has possessed from the beginning of his career. He has not denied himself one of the real blessings of life. All the healthful and satisfying delights of labor he has enjoyed without many of its anxieties. What a zest it gives to his labor that he is improving and embellishing what is to him, and what will be to his children, a beloved home."

As appropriate to the subject in hand, a writer, under the signature of "B," contributed the following sensible remarks:

"A farm should be the home, and its management the business of the owner. It is true, one may be hired or worked on shares, but very seldom do we see land, cultivated under such circumstances, managed in a way worthy of the name of farming. Ownership seems necessary to a proper appreciation of the characteristics and powers of the soil. We again see a movement in the real estate market—sales and purchases of farms, and it suggests some thoughts on what one should look to and seek for in buying a farm.

"Considered as the homestead and abiding place of the owner, a farm should be pleasantly and conveniently situated. The health, comfort, and happiness of those who occupy it, are of the first importance; so every social and physical influence which bear upon them should have due weight in determining a choice. A healthy locality should be considered far above a fertile soil. The thousand things which promote home-comfort will compensate for many pecuniary disadvantages. Happiness, the enjoyment of social privileges and blessings, go far to make a sterile soil of greater value than the most productive, where a moral miasma prevails. A situation of easy access to the great routes of business and mails, with educational and religious privileges of a high class, would be considered of the highest importance by the intelligent and cultivated man, who would enjoy the best privileges of American life and society.

"Another thought. The new location should be suited to the tastes and character of the purchaser. Men of mature age are usually of fixed habits and dispositions, such as do not change with a removal to another home. They should find then, in the new, the best pleasures and conveniences of the old, and as many improvements as may be. But if circumstances require any considerable change, it should be remembered that to make it will require some exertion and energy—they must expect this, or meet disappointment. Their children may find a happier and better life in the new locality—the sacrifice of old habits can be made for their sakes.

"As a business, the requisites of successful farming depend to a considerable extent on the choice of the farm. It should be one which the owner has the means and the understanding to manage. One cannot put all his capital in land, and expect to farm profitably on credit and makeshifts—often so cramped that all improvements are out of his reach. As well might the merchant put his whole capital into a fine store, reserving nothing to purchase the goods wherewith to fill the shelves and attract customers. It requires as much capital to stock and carry on a farm generally, as to pay for the land itself. The farmer needs capital to keep his credit good—to take advantage of the markets in buying and selling, and in making seasonable improvements. A farmer loses money who is compelled by want of money to sell his crop at the lowest stage of the market, or who cannot command extra labor in any emergency of the season, or who is obliged to wait for years to get a few hundred dollars to drain a swamp that would pay him the interest on a thousand dollars as soon as the work was done.

"The farm should be suited to the products which it is desired to devote it to. The taste and experience of the owner will incite him to undertake certain branches of farming, but some soils are best calculated for grain-growing, others will produce extra fruit, others have grass and water for the dairy, or stock generally, while occasional locations are to be found where all these may be combined to a greater or less extent. These things should be taken into account in buying a farm.

"Then market facilities are to be considered. In the management of a farm much depends on this, and it is a matter of moment whether it will cost five cents or fifty to bring a dollar's worth of produce to the consumer. In the vicinity of large towns the production of garden crops is often very profitable, while at a distance from market no dependence can be put on such products. The one can grow a large variety to dispose of—something every week bringing in the cash—while the other must necessarily devote himself to a few leading articles, his harvest occurring but two or three times a year. But the recent great increase in the means of transportation has done much to equalize the value of farming lands throughout the country, especially when devoted to the more valuable and least bulky articles of produce.

"Again, a farm should possess in itself good capacity of production, so that it may be readily and profitably managed, in such a way as to retain and increase the fertility of the soil. A farm easily worn out—a course of management rapidly exhausting the fertility of the soil, would soon bankrupt the farmer; his business would no longer be remunerative; his home and his comforts would soon pass away. Hence it is not all to buy a farm; one must also have the skill to manage it rightly. To do business profitably, one must understand business principles and carry them out, and nowhere is this more important than upon the farm. The question is often debated whether farming is really profitable or not; but, could we only see the fortunes lost by the careless habits of those who pursue it, the decision would soon be arrived at."

These remarks drew "F." once more into the arena, in the following communication:

"B. has some very good remarks on buying a farm, with the most of which I fully agree. But the general tenor of the fourth paragraph is not of that general practical character that is best calculated to benefit the great mass of American farmers. True, it looks and sounds well on paper, and could we have farmers made or got up to order, giving to each one the amount of land or capital he is or may be capable of farming or using to the best advantage, it would be all correct. But this is not the case. Instead of having sufficient capital to buy stock and carry on their farms to the best advantage, the great mass of farmers have had to make and get together the principal part of this very capital by farming. Consequently it is of little use to tell them what sized farm can be carried on to the best advantage, or the amount of capital it is necessary to have to commence farming with, or to carry on the farm after it is purchased; for, as a general thing they have to buy, calculate, and manage, according to the circumstances in which they are placed.

"Inseparably connected with this question, is the one of going in debt for land, or of paying part down, and the balance in a term of years, the money to be made out of the use of the farm. Although there is much in the accounts of individual experience published in agricultural papers, that goes to show that men have done well, and have been very successful, when in debt, yet it is the general custom of editors, and more or less of other writers, when treating these questions in a general way to strongly oppose—as is done in the paragraph above mentioned—any running in debt for land. The idea appears to be—and in the abstract it is well enough—that a debt on a farm is a clog, a hindrance, a something that stands in the way of or prevents the practice of the most successful course of farming—an idea that seems to view the farm, and the capital necessary to carry it on, as an end already attained, and only as a means to make a living, and accumulate more capital. Yet, with the great majority of American farmers, farming has been the principal or only means to get the farm, as well as the stock and capital to carry it on. Hence, on commencing, they are not called to decide on how large a farm, or how much capital they have to farm to the best advantage; or whether they can do as well when in debt for their farms as they could if out of debt, with plenty of money to use; but, on the contrary, the great practical question is how to get the farm. Comparatively few have the farm, or the money to buy it furnished, to begin with; and, though there are some that, having sold a farm, have money to buy another with, though, as a general thing, such men have sold with a view of improving their circumstances, aiming or intending to buy a larger or better farm, which will cost more than the one sold, and in most instances put them in debt; still, as a general thing, it was the same with them as with others at the first commencement.

"Hence, I repeat, the question is how to get the farm. And involved in this question are many others, such as, is it best to try to get a farm by buying and running in debt for the whole of it? or is it best to wait until money enough is made in some other way to pay all down? or is it best to pay part down, and run in debt for the rest? Or, to put the question in a different and more practical shape, if a man has only money enough to buy and pay down for twenty-five acres, had he better buy fifty acres, running in debt for twenty-five acres; or, if he had only money enough to buy and stock fifty acres, had he better buy that amount, or buy one hundred acres, running in debt for fifty, and arranging and managing his stock so as to raise enough in a few years to fully stock his farm? Or had he better take the advice often given in the papers, that is, only buy when he is able to pay for land enough to farm to the best advantage, and to have an amount of capital equal to the cost of the farm, as advised by B.?

"Now, it would be useless to give any farmer such advice here in this section of the State. Good farms sell here at from sixty to eighty dollars an acre, and the best size for a farm, to work to the best advantage, is probably from one hundred and fifty to three hundred acres. Such farms as well situated for, and as well adapted to farming to the best advantage, as seems to be demanded by the general tenor of B.'s remarks, would cost from $10,000 to $20,000, to say nothing of adding as much more to stock and carry on the farm. So that in almost all cases to advise a man not to commence farming until he has such a farm, with such an amount of capital to work it with, would be equivalent to advising him to not buy at all. It will do little good to show him that farming can be made the most productive and profitable where the farms are large enough to use all the most approved tools, implements, and machines to the best advantage, and where there is all of the capital that can be made available or used profitably; or to tell him that the greatest prosperity of the community and country requires, and the character and standing of the farming class may be promoted by this kind of farming. He will, perhaps, assent to all you say, but tell you that this is not the main question with him. That the main object being to get a farm, and having money enough to pay from one-fourth to one-half down for such a farm as he wants or would be glad to get, with sufficient means to get the necessary teams, tools, etc., to work it to good advantage, and to provide for supplying or raising all necessary stock, the question is, is it best to buy? Can he pay for the farm, and is there any better course for him to take than to buy and go to work and pay for a farm?

"Then again: You may tell him that some of those that run in debt for farms fail, and lose all they have; while others have a long, hard, up-hill business in paying for their farms. But he may answer that it need not necessarily be so; that more or less fail in all kinds of business, situations, and circumstances; but that, as a general rule, those that run in debt more or less for their farms succeed in paying for them, and often attain to very comfortable and independent circumstances, without suffering any very great hardships or privations. And did he live in this section he might say, that as a general thing all of the best and ablest farmers are men that, having begun with but little, if any, capital, have made the principal part of their property while in debt; that there are very few farmers that are considered forehanded, or well off, but that have at some time been in debt for farms; that some of the ablest men, worth from $20,000 to $25,000, have not only commenced farming with very little if any capital, but doing business on the general principle of buying and running in debt in most part for land, and then going to work and paying for it; when one piece is paid for, soon buying another, and so keeping on, seldom out of debt long at a time, but generally making money, and satisfied and contented, they have eventually attained to circumstances that would be considered very satisfactory to most men.

"True, their course or system of farming, though generally not bad, is not up to the highest standard of modern agriculture. But this is owing more to following old practices and prejudices, and the kind of, or want of education, than the want of means. And I may also say that their general course of farming is being gradually but surely and largely advanced and improved by the younger class of farmers that are coming on; some of them having, in proportion to their chances and opportunities, been more successful than any of the older class.

"A few words in regard to running into debt for land. A land debt, when entered into with proper precautions, and well guarded,—the farm being worth considerable more than the debt, and in a fertile condition, giving reasonable assurance of producing more than enough to pay the interest and a portion of the principal each year, and payments arranged so as to give ample time to meet them,—is not a thing to be so much dreaded. In fact, it often seems to be an advantage. It seems to furnish an object, an end to strive for, to be attained to. It gives the farmer a home, a something that he wants and means to keep, that he would be sorry and ashamed to lose, and altogether something that will induce him to work harder and manage better than he has ever done before, rather than fail in. This seems to be so well understood among farmers that the remark is often heard, 'that the best way to make money by farming is to buy and run in debt for land, and then go to work and pay for it.'

"Perhaps I cannot better close this part of the subject and this article, than by giving a little of the experience of a farmer in the north part of this county, as related by himself to a friend of the writer. He said, that having got his farm paid for, and a little money ahead, his family thought they must have something a little better than common, so they persuaded him to buy a $300 carriage. When they

got this home they discovered that they had no harness to match, so he had to buy a costly harness. Then they said the old farm-horses were not fit to go before the new carriage, so he had to pay $300 for a better pair. And then it was fine clothes and costly furniture, etc., etc., until, as he said, he could stand it no longer. And he continued: 'I have now drawn to market my last load of wheat, and have a thousand dollars in money. I can buy a neighbor's farm that is handy and convenient to mine, for $5,000. I think it is worth it, and I am going to buy it to-day. I find that, as things are going, about all we can earn will go to support extravagance; while I know if I go in debt for a desirable piece of property, like this farm, my family will go to work and help me pay for it. We will then have something really valuable to show for our labor, and that some day will do my children some good.'"

With the following brief rejoinder by "B." the discussion was closed:

"The question of running in debt for a farm, under certain circumstances, is argued very fairly by F., of Orleans county, N. Y., in the *Country Gentleman* of March 26; but we see very little in the paragraph of B.'s article to which he refers, to call out his remarks. The advice is given not to put all one's capital in land, retaining nothing to stock and carry it on; and something must always be reserved for this purpose. No business can be carried on without capital of some kind—because something can never result from nothing—valuable crops always require seed and culture.

"In buying a farm the mistake generally made is in not providing sufficient means to carry it on. There is no difficulty in getting extended credit for a part of the purchase-money on a good farm, and we believe the most direct means of becoming in fact the owner of one, is to go in debt,

if necessary, to a sufficient amount to provide ample means for its stocking, cultivation, and improvement. To make money rapidly, by farming, requires ample capital; those who cannot command it, lose, to say the least, one-half the profit they would otherwise secure. It may be true that those farmers who can see no better investment for their gains than the purchase of finery, need the pressure of debt to forward their advancement in property. In many cases, however, it would be more profitable than any other course, to spend their surplus capital in improvements on the old farm for several years, before adding a single acre to its extent in surface.

"We have seen and known of so many losses in farming, for the want of capital, that it makes us the more particular to urge this matter on the attention of farmers. We hope F., with his wide extent of practical experience, may have seen something corroborative of this view of the matter, and that he will present it and illustrate it with his usual force and ability, in your columns."

The remarkable discussion thus quoted will not fail to give the reader many new and practical ideas on the subject of getting a farm. It ranges over many branches of the question, and contains a mass of robust good sense which an ambitious young man cannot too closely study. The suggestions made are not those of enthusiastic theorists, but the fruits of long and grave experience, such as, having been in most cases successful, are entitled to the highest respect. It is to be noted, moreover, that the question of how to get a farm being once started by one who saw the difficulty of obtaining it without capital, the subject was immediately recognized to be one of great and general importance, and that a large number of intelligent and experienced men hastened to engage in publicly discussing it. As in a multitude of counsellors there is sometimes wisdom, so in this friendly collision of opinion the seeker after knowledge on the subject will obtain new light, strong encouragement, and every reasonable incentive to induce a beginning. This country must contain thousands of men who, in the various ways thus indicated, are now living on homesteads which their fathers or themselves have thus acquired.

It was the reading of this discussion in the columns of the *Country Gentleman* that led me to prepare this volume. It struck me that all that ought to be said on the subject had not been discussed. Indeed, the columns of a weekly journal do not afford space for a consideration of the numerous aspects in which it might be presented. It was evident, from the number of writers who volunteered to throw light on the question, that it was one in which a general interest must be felt. Hence my effort to increase the usefulness of the discussion by grouping together all that has been already said, with more that has been omitted.

CHAPTER V.

Exhausted Farms always to be had—Thriving Tenants—Owners anxious to sell—Bartering Farms—A lucky Beginner—City Owners—Taking Advice—Where to Search—Saving a poor Farm—Struggling with limited Means—A Cry from a Working Man.

THE discussions quoted in the preceding chapters, though very clear and full, are far from exhausting the subject. There are multitudes of farms in all the Atlantic States whose owners have never contemplated working them. Some purchased as an investment, thinking farm land the safest property to hold. They rented them for a succession of years to tenants who skinned them with merciless assiduity, making them a heavier burthen to their owners the longer they held them. Year by year they thus became poorer and less productive. If rented on shares, as is generally the case, the tenants appropriated most of the product, the owners getting little or nothing. The latter lived away off in some distant city; they rarely visited their country property; they had neither taste nor opportunity for seeing whether it was handled wisely or honestly; the tenants were thus left to exhaust the land as rapidly as they could, and were depended on to make report, at the year's end, of what was generally a bad season, of poor crops, poorer prices, and a still poorer return to the expectant owners.

If the plundering tenant were discharged, he was generally succeeded by another whose genius for stealing was superior to that of his predecessor, as upon an exhausted farm he must skin more severely and steal more largely, to obtain a living. If the first thief took nearly all, the second was sure to take what was left. The owner being thus annually robbed, becomes tired of what he once considered the safest investment, and is anxious to sell.

But a farm thus long the victim of spoliation, finds few cash buyers. It will require time, labor, and money, to restore it. Depreciation is a rapid process, but restoration a slow one. Cash buyers prefer land in good condition, considering it cheaper to enter on a farm in prime order, at a high price, than on a poor one at a low figure. It is fortunate for the country that all who are looking for farms do not entertain the same opinion. If they did, the numerous tracts which have been thus skinned to death would be reoccupied by the forest, as none would be found courageous enough to undertake the slow and costly process of resuscitating them.

The disheartened owner thus finds neither a buyer nor an acceptable tenant, nor is he so situated as to give his farm the least attention himself. He does not need the purchase-money—all he desires is to find some reliable man to relieve him of the care of an intolerable burden, by taking the farm on some terms. It must be occupied by somebody, as absolute desertion of house and grounds is as ruinous to a farm as idleness to the machinery of a cotton-mill. Should a young and competent man now offer himself as purchaser, fortified by good character and a knowledge of his calling, the chances are that he can secure the farm on such terms as to price and payment, as will make it the great and successful operation of his life. Such a man, having constantly looked forward to beginning for himself, will have saved a few hundred dollars; and these will be found sufficient for a start. Others have begun in like circumstances, with no capital but their hands, and have succeeded.

I have seen more than one farm thus owned, thus plundered and exhausted, thus an encumbrance on the owner's hands, and thus passing into the possession of men with little or no capital, who in a few years had paid, from its own products, every dollar of the purchase money. From the day they entered into possession they enjoyed the comforts of a home. It might at first be scanty, rough, and inconvenient, but still it was home. Every tree they planted became an investment for their own exclusive benefit. In every furrow they turned, some golden particles were discoverable at the bottom. Every spoonful of manure they bought or manufactured, was equivalent to a fund invested at more than compound interest.

There are hundreds of poor farms now held by their owners because no buyers can be found. Men in search of land should seek them out, bargain for them at low prices and at long terms of payment, and enter into possession. Let no want of capital operate as a discouragement, but go resolutely to work. Such beginners should avoid great farms. Far better to begin with thirty or fifty acres, pay for that, and then, if more land be indispensable to comfort, enlarge the boundary.

There are two other classes of owners on whose hands farm property hangs either lightly or with oppressive weight. In all the large cities there is maintained an active trading business, in which houses, lots, land, merchandise, and patent-rights, are passed rapidly from hand to hand. Money is sometimes mentioned, but rarely paid—the whole transaction is one of barter. One who will take the pains to look over the registers kept by these city dealers, in which are entered the properties they have for sale or barter, will be astonished at the extent and variety of the collection. There is almost every thing that anybody can desire—houses, lots, farms, mills, factories, water powers, wild land, some of which is within an hour's ride of a great cash market, and others two thousand miles away. Many of the farms have been taken in barter by city owners, whose sole business it is to get rid of them as quickly as they bought them. In some cases money is wanted, in others it is not, the barter being repeated by exchanging the farm for something considered more salable. Thus the barter is kept moving until some commodity turns up which can be converted into money.

I saw at one of these agencies, in Philadelphia, a tolerably good farm sold in exchange for a half interest in a patent pump, which was subsequently found to be a failure. On another occasion, a young man of six-and-twenty, dressed in a farmer's everyday suit, came in to look after a place of seventy acres which he had seen advertised. The advertisement did not state exactly where it lay, but the low price of $700 attracted his attention. The seller opened his thick book and read out a minute description, on hearing which the inquirer immediately recognized the farm as one of which he had long had some knowledge. After a very short parley he bought it. This farm was improved with buildings which had cost $1,200, though now old and out of repair. The fencing could not have been placed there for less than $300, while there were other appliances about the house sufficient to make a moderate family comfortable. There was wood enough on it for fuel, and it was within two hours' ride of Philadelphia by railroad.

But why should it have been sold so low? The former owner was a lazy, thriftless fellow, and, like the weeds on his land, had fairly gone to seed. Poor, of course, the longer he remained there the poorer he became, and thinking he could better himself in the West, where land was cheap, bartered off his farm for a half section, 320 acres, of Missouri land. This half section had cost the buyer $500—at least he had taken it at that figure in a previous trade. He had never seen it, neither had he ever been to examine the farm for which he exchanged it. His business was to buy and sell, not to examine property or to keep it. Thus $700 was to him a capital price, the more so if it should be paid in cash.

The lucky young man who bought it had been for six years a careful saver of his earnings, and had $800 in hand. He had concluded to marry and get a farm. His intended wife, also brought up in the country, had saved $200. The union of these two little capitals thus gave him the very start in life he was seeking. But his excellent character was good for another thousand, whenever he chose to borrow. Buying such a property at such a price, and occupying and working it himself, must have laid for him the foundation of a certain independence. This incident forcibly illustrates the value of even small savings—how they sometimes enable a deserving man to seize upon the golden opportunity the moment it presents itself.

There is another class of city owners, not professional traders in property, who, having something which they were anxious to part with, have exchanged it for a farm, thinking thus to better themselves. But these soon discover that a farm so far off that one can rarely see it, is a great plague, and speedily become anxious to sell, even at a loss. If a sale for money be found impossible, then one on credit is gladly made. The main object is not so much to get money as to shake off a perpetual care. They discover that an idle farm goes to ruin as rapidly as an idle steamboat.

Here are different classes of persons, all owners of farms, and all governed by the same feeling, that of anxiety to get rid of them. These reside in cities.

It is apparent, then, that one likely way to get a farm is, in the first place to seek out such properties as are known to belong to distant owners, and which are at the same time being skinned by worthless or dishonest tenants; to obtain all the information possible in reference to the property; and then to find the owner and negotiate with him on the basis of a purchase, with ample time for payment. He will listen more respectfully to the man who proposes to buy than to one who merely proposes to rent. An offer to buy implies ability to pay—at least some time—and holds out a prospect of the owner being relieved of a great annoyance. Renting is synonymous with continuance of an old and hateful grievance. Small means will be no hindrance against the application of a worthy man in these circumstances. Character will be the preponderating ingredient towards success, and, as in most others, will determine the question in his favor.

I could recite three instances where this advice was given, and, being acted on, success attended each application. In two of them, the good conduct of the applicants was so conspicuous as to secure the confidence and friendship of the men from whom they had purchased, to such an extent, that, whatever facilities were subsequently needed, were voluntarily supplied. Character and conduct secured them friends, and were absolute equivalents to capital. Kind words, moreover, were constantly spoken. Though cheap, yet they were valuable because inspiriting. Some men are naturally given to them; feeling that such utterance is like lighting another man's candle with your own, which loses none of its brilliancy by what the other gains.

So, too, the shrewd man with few or no dollars in his pocket, should watch the advertising columns of every newspaper he can lay hands on. Wise men resort to them to let the world know what they have to sell, and other wise ones—the buyers—drink daily at the same fountains of intelligence. It is notorious that vast fortunes have been made by those who have freely advertised their wares. Is it not reasonable to presume that those who read and purchased have had their share of profit? There are more bargains to be found in the newspapers than at the auctions.

He should also consult the thick registry books, kept by the numerous dealers in real estate, in all the large cities. These books contain descriptions of thousands of properties on sale. Many of the latter will be found to be desirable bargains. He must be hard indeed to please, if unable to find among them some one to exactly suit his wants. Having thus pleased himself, the chances are that he will find little difficulty in being able to please the owner. It may be urged that farmers, especially young and inexperienced ones, are not business men enough to undertake a pilgrimage of this kind to strange places, among strange men. But bargains do not come to you ready made. The gold of California did not come to those who remained at home, but to those who went after it. They must break away from the old routine of their lives, remembering that our customs and habits are like ruts in the roads. The wheels of life settle into them, and we jog along through the mire, because it is too much trouble to get out of them.

I could point to one farm thus taken from an owner so disheartened that at one time it was seriously contemplated to abandon it. The soil seemed hopelessly sterile, was chiefly yellow dirt with gravel, and had apparently as little capacity for retaining manure as a sieve for holding water. But deep ploughing and heavy manuring with fertilizers, manufactured on the premises by hogs and cattle, have brought it up to yielding thirty bushels of wheat per acre. Patience and perseverance, with very moderate capital to begin with, overcame every thing. As may be supposed, it was purchased for a small sum, but in ten years could have been sold for a large one. Instances of recuperation are too numerous to be recited.

The design and object of this volume cannot be better promoted than by quoting from the *Country Gentleman* the following inquiry, made by one of the large class of persons who are constantly looking for a country home, with the editor's answer. His ripe experience of agricultural subjects gives the highest value to his remarks. The inquirer describes himself as a young lawyer—

"Making about $400 per year by my profession, and with almost no hope of increase, as the market is overstocked; and I am sensible that I am neither a Kent, an Emmet, nor a Story. Having a wife and child to support, I am anxious to try some other business, that will better enable me to do so than at present. Would farming be better?

"I possess a thorough theoretical knowledge of agriculture, having at college made chemistry and botany my particular studies, and have, for years, been a constant reader of the best agricultural books and periodicals, the *Cultivator* included. I have had considerable experience in gardening, and have been very successful. There my qualifications end. I have never worked a whole day in the field, and am too feeble in body to do much labor of any kind.

"About five minutes walk from my house and from the town of B——, there is a piece of land containing sixteen acres of good clayey loam, sloping toward the south about one hundred feet down to the seashore, where marine weeds are abundant for manure. It is well fenced, but the buildings are worth nothing. It has been badly cultivated, without any kind of manuring. It rents for $50 a year, and may be bought for $1,000, on easy terms, say $100 cash, and the rest in eight years. My whole available funds are about $200. Would it be prudent, then, to buy it? I can hire a good man for $80 a year and board. Produce brings fair prices, and is readily bought up: barley, at $1 per bushel; oats, 60 cents; potatoes, 60 cents; Swede turnips, about 40 cents; and hay $15 a ton, this year, and about $9, in other years. Indian corn we rarely grow, being liable to early frosts. Will you give me your advice? It would not be necessary to relinquish the law altogether; I could probably make $200 by it, and still work the land."

The case is stated with lawyer-like precision, whereupon the editor replies in the following language:

"In giving advice, in such a case as this, it should be borne in mind that more depends on THE MAN than on the nature of *the business*, provided the latter is such as to give an opportunity for the exercise of the energies. We have known of more than one instance where two men, with similar opportunities and means, have succeeded very differently at the same business—one failing, and the other accumulating wealth. Hence we cannot advise, in a general way, the purchase of a farm, by running almost wholly in debt for it. A few would easily work out; but to most, the debt would be likely to prove a long-continued and oppressive load. It must depend upon the management, tact, and *economy in every sense of the word*, possessed by the purchaser in question.

"If we were to give *one rule* in business for beginners, which we should place at the head of all others, it would be—FEEL YOUR WAY. Do not undertake any thing untried, on a large scale, no matter how promising the results may appear. The most uniformly successful men in business have nearly always pursued this course; and we could, on the other hand, name many instances where large and bad failures have resulted from a different practice.

"Our correspondent should not give up his practice of law immediately. He must depend on that *mainly* for two or three years at least, until he gets under way in farming. If he could rent the land for two years, with the privilege of purchasing, it would undoubtedly be best. But this, probably, cannot be accomplished. He must therefore take into consideration the probable cost, in addition to the land, of animals to stock it (for even the smallest farm should have some animals), the expense of a horse or of a yoke of oxen, of the various necessary implements to work it to the best advantage, and of proper buildings. A man must be also hired to do most of the labor in the present instance. All these will be found to consume more than the proceeds of the land for the first year or two. If, after all these calculations, he can be sure of meeting his interest and other payments, allowing for disasters and contingencies, a purchase may be made.

"Great judgment and skill should be exercised in selecting *first*, those improvements which promise the greatest returns with the least outlay. On this branch of the subject a large book might be written. We can only say here, make a list of *proposed* improvements—examine from all the *practical* knowledge that can be collected, and to some extent from limited experiment on the spot, the probable cost on the one hand, and the probable advantages on the other, and then select *first*, those showing the largest percentage of profits. These may be some kinds of manuring, or some cheap and efficient underdraining, or deep plowing of the tillage land, or heavy seeding of the grass land, or the cultivation of certain crops, or the introduction of certain animals, remembering the rule in all cases, FEEL YOUR WAY.

"Sixteen acres constitute but a small farm, but such a farm, skilfully managed, may after a while be made to produce a considerable amount. Ordinary field crops alone will not be likely to produce even one-half of the $400 now made by law practice; but these, with fruit raising, rearing fine animals, producing marketable garden crops, and perhaps the more salable fruit trees, may, with *skilful hands*, be made to increase the product to an almost indefinite extent.

"We have heard wealthy farmers assert, that not over two per cent. on the cost of their farms and the capital to stock them, could be fairly relied on as an average for all seasons. But this estimate is made for ordinary superficial farming. We have found, by experience, that a better mode of practice, economically pursued, would, without any special trouble, double the products, and triple or quadruple the net profits. For instance, where a ton and a half of grass are commonly cut, four tons have been produced simply by heavy seeding and plastering; where thirty bushels of corn have been the common crop, seventy bushels have resulted from well applied manure, selected seed, and good cultivation; and where only two hundred bushels of carrots or rutabagas were ordinarily yielded, six to eight hundred have been obtained by performing every part of the operation promptly, in the best manner, and on a deep and rich soil. We have known a gain of more than one hundred dollars a year on a single farm from a selection of the most efficient tools, and proper labor-saving machines. We could also name some farmers, who, instead of reaping an average net profit of only two per cent., make at least twenty per cent.; and some of the best farmers of Western New York (and doubtless elsewhere) *clear* from $700 to $900 from every hundred acres—and in one case about $6000 have been made in a single season from a five-hundred-acre farm. The owners and managers of these farms were active, intelligent, and energetic men, of long experience, always in the midst of every important operation, and we need scarcely add, constant readers of the best agricultural publications of the day."

Such are the opinions of one who has long been regarded as an agricultural authority. They are cautiously expressed, but are not discouraging. They strengthen the position taken throughout these pages, that it is not the mere land which makes the cultivator independent, but the skill and industry with which he handles it. They may be said to lie at the foundation of all intelligent, well-directed labor. An animal will grow if generously fed; if not, he must remain nearly stationary. If not fed at all, he will assuredly die of starvation. It is precisely so with agriculture. Crops must be fed, or they cannot be produced. In this case, the feeding consists of thorough tillage and manure. Both are costly items, but they are indispensable to success. The great question is how to command them in the largest abundance at the smallest cost.

It is granted, on all hands, that the soil should be saturated with manure. Mere hard work will not procure it in the necessary quantity. Brains must be made to come in as the leading helper. Hence intelligence is necessary—the mere land will be found powerless to do all that too many expect it to accomplish. Women, sick men, even absolute cripples, unable to perform an hour's work, have been highly successful farmers. But if deprived of hands to work, they had heads to superintend.

Just before the crash of 1857 came with such desolating severity upon the country, a working man, a resident of the city of New York, who had somehow scraped together $4000, found himself precluded from increasing it by high rents, dear food, and the inevitable increase of expenses on every side. In this dilemma, he addressed himself to the *Tribune*, in these words:

"I want to know what chance a man would stand in the country to take up farming—not out in Kansas—but say in Jersey, or the western part of this State, or out West near some improving location, so as to get away from high rents and dirt, and breathe a mouthful of fresh air—say a little place of fifty or sixty acres, with house and barn, one horse, three or four cows, a few sheep, some poultry, all paid for and clear of debt, so as to have $1,000 or $1,500 at interest, just to have a little loose change coming in. Could a man live better, feel better, and be less a drudge, than to stay here in New York, and drudge, drudge, from day to day, and month to month!"

What a revelation is here given of the workings of one mind among the large class for whose information this volume has been written! To this string of questions the editor made the following pointed reply:

"Farming is a vocation, requiring knowledge, experience, and skill, like any other. No man born and reared in the city can remove to a farm, at thirty or forty years of age, and become immediately an efficient, thrifty, successful farmer. He will have much to learn and something to unlearn; and if he should get through his first year of farming without using up $500 of his capital, he many consider that he has done well. Yet if he will keep his eyes open, take counsel from his neighbors, take two or three good agricultural papers, and read them carefully, we believe he can render himself a fair average farmer the second year, and something better than this thereafter. But, Is such a change as this desirable?

"We answer, Yes. If, with average capacities, and a capital of $4,000, you can, by steady industry, make nothing beyond a bare living in the city, we hold that you can do better in the country. If you buy your land for its fair valuation (and a great deal in this quarter is held twenty to fifty per cent. above that mark), and use it well, it must be steadily increasing in value. Your buildings also will necessarily be enlarged and improved—this year a corn-crib will be added, next year an ice-house, and so on—and though your surplus funds will rather diminish than increase, and you will hardly see a dollar where you now see ten, you will be insensibly crawling toward a competence. Your children will every year become more and more helpful to you; your young fruit-trees will come into bearing; and you will have at least twice the produce to turn off six or seven years hence, that you can spare the first year.

"As to drudgery,—a man who aspires to earn a good living and rear and educate a family by honest, straightforward work, must be diligent as well as frugal,—there is no help for it. Still, a farmer with $4,000 capital well invested, or even half that sum, need not labor excessively. From April to November, he must work steadily and energetically at least ten hours per day; but in winter he may moderate his exertions and give himself a week or more to visit relatives or friends, when he sees fit. Farmers do not work so many hours, on the average, as do the mechanics of this city.

"As to location; we think a man with $4,000 may buy a fair farm, in New Jersey, or one of our river counties, stock it fairly, and have $500 over for lee-way; but he can, of course, buy a much larger farm and stock it much better from such a capital in a newer region, or even in western New York. Every section has its advantages and disadvantages; we should more strenuously insist on a healthy locality, than on fertile soil. With a cash capital of $4,000, you will be a rich man in almost any new settlement; but frontier life has its inconveniences, even for men; still more for women and children. Investigate and decide for yourself."

CHAPTER VI.

Wanting the Best—The Poorer Lands first Cultivated, then the Richer Ones—Value of Swamps—History of three of them—Cranberry Swamps of New Jersey—Power of Example—The Mississippi Swamp Interest—Wealth following Reclamation—Public Loans to aid Drainage—John Johnston, the Great American Tile Drainer.

It is a feature of American thought and habit to be rigid and exacting. We are too apt to reject the moderately valuable, and to insist on having only the best. The habit has infected even the children, beggars though some of them may be. A ragged little urchin came one morning to a gentleman's door, asking for old clothes. He brought him a vest and a pair of pants, which promised to make a comfortable fit. Young America took the garments in his hand, and examining them as closely as if he had been buying them, then, with a disconsolate look, exclaimed, "There ain't no watch-pocket!" Now, the truly great are humble; just as those ears of corn, and those boughs of trees which are the most heavily laden, are seen to bend the lowest. The urchin's impudence represents the national propensity—we must have the best.

But what we may conceive to be the best for us, frequently turns out otherwise. It is thus in seeking to obtain a farm. He who is looking exclusively for upland and meadow, all ready for the plough or scythe, will be startled should he be told that he can do better by buying a swamp. The reader himself will probably be staggered by the proposition. It is a new one, truly, but will justify examination and analysis.

More than fifty years ago, Mr. Ricardo communicated to the world his discovery of how men came to pay rent, and others to exact it. His work was accepted by political economists as a text-book; his views were assented to without dispute; and for forty years no one ventured to doubt their correctness. His theory was a very simple one. He laid it down as law, that in the commencement of cultivation, when population is small and land abundant, the best soils, such as yield the largest returns, are the only ones cultivated; that as population increases, land becomes less abundant, creating a necessity for cultivating a less productive quality, when, as population again increases, resort is had to still more inferior land, then to a third, and afterward to a fourth class of soils.

These propositions, with the theories established on them, were for many years accepted throughout Europe, as well as in this country, as undeniable. But our distinguished political economist, Mr. Henry C. Carey, in his "Past, Present, and Future," published in 1848, has demonstrated their entire falsity. The foundation thus knocked away, the fabric of theory which Ricardo had reared upon it fell to the ground. Mr. Carey proved that the facts were exactly the reverse—that the land which man cultivates in the beginning is of the poorer qualities, and that that which he last brings under tillage is invariably the best.

Take our own country as an illustration, because all can judge from facts that are occurring everywhere around them. Here, the settler invariably begins by occupying the high and thin lands, which require little clearing and no drainage. Such yield him but moderate returns for his labor. But in time, as population and wealth increase, he travels down the hills, clearing up as he goes, until the bottom has been reached. There the hill terminates, and there the meadow or the swamp begins. It contains the rich deposits which for centuries have been washing from the hilltop and the hillside, the loss of which had thinned the poorer soil into which he first struck his spade.

Every reader knows that no settler begins his clearing either in swamp or meadow, yet every one is aware that in such spots the richest land is to be found. He knows, moreover, that lowlands are the last to be reclaimed. To bring them into tillage requires money, skill, and time,—courage, also, may be mentioned as an indispensable auxiliary,—the courage to undertake a task which a succession of owners had carefully avoided. Ditching is familiar to most of us, and can be cheaply done; but thorough underdraining was comparatively unknown among us until within a few years; and it is only by resorting to it that the lowland can be effectually reclaimed. In a primitive condition of society, therefore, with money scarce and land abundant, the richer soils are neglected for the poorer.

The early settlers, throughout New England, established themselves on the higher lands along the river courses, leaving to their more wealthy successors the task of clearing up and draining the swamps. New York and New Jersey were settled on the same plan; the higher grounds were first occupied, while vast meadows of extreme richness were neglected, because they required drainage. New Jersey contains multitudes of abandoned clearings, made by the first settlers on the poorer soils, but long since deserted for the richer ones. Maryland abounds with evidences of the poverty of the soils first occupied, and of the richness of the meadow-farms subsequently brought under tillage. In Pennsylvania, the oldest habitations were always the most distant from the rivers. The rule prevails throughout the South and West. The higher and drier lands, in Mississippi, were peopled first; the rich bottom lands of her great river were subsequently reclaimed by ditches, and the vast embankment which now keeps out the annual overflow. Throughout South America the same extraordinary uniformity of practice has prevailed. In England, over the European continent, and wherever man has found a foothold, no departure from it can be discovered. The law laid down by Mr. Carey may thus be regarded as incontrovertible—the poorer lands of a country are settled first, the richer ones are settled last.

It remains for me to apply it to the subject in hand. I have intimated that a man desirous of obtaining upland and meadow, ready for the plough and the scythe, would be incredulous if told that he could do better by purchasing a swamp. Unless he comprehended the foregoing law, or had seen it illustrated, or had had it clearly explained to him, it is natural that he should be confounded at the proposition. It would seem to him like asking for bread and receiving a stone. But his object is not only to get a farm, but to get the best one, and at the lowest price.

Now, we know that the best land is always to be found in the swamp. There grows the heaviest timber, there vegetation shoots up with the rankest luxuriance, there the dark mould, which we call soil, is uniformly the deepest. It has been accumulating there for ages. Rains have washed down upon it the rich soil of the surrounding hills, for centuries before the white man had trodden them. The forests that covered them have showered upon it their annual wealth of leaves; and winds have blown to it, from other woods, additional stores of foliage. Decay of the living and the dead has been going on without interruption. The original depression has become filled, many feet in depth, with a deposit so rich that the owner sometimes spreads it over his grounds, half suspecting it to be manure. It is, in fact, a mass of fertilizing matter of uncounted value. No intelligent man can doubt it.

But the owner is ignorant or neglectful. He considers it only as a swamp. In his family it has had no other name. His boys hunt in it for skunks or rabbits, or for birds or honeycombs. It is known through the neighborhood as the swamp. It is good only for an occasional load of rails, perhaps a load of wood. Though assessed as waste land, yet it does not yield the annual taxes. Such it is considered by all who know it; and such, while in this condition, it really is. In vain has the owner endeavored to dispose of it at a low price; it is too well known, and too little valued by those who do not understand the capabilities of ground thus situated, to tempt them to buy.

Yet this swamp may contain ten acres or a hundred, and be so located that a single ditch cut through the centre will render it comparatively dry. Cross drains on both sides of and emptying into the central ditch, composed of wood or tile, if sunk at intervals of thirty feet apart, will render it firm enough for the plough. By covering them they become underdrains. The main ditch itself may in some cases be covered also, leaving the ground, when cleared, without break or obstruction. It needs no manure, for nature has there concentrated an untold wealth of her choicest fertilizers. Who can doubt that fifty acres of a swamp like this, bought at a nominal price, and thus treated, will yield to the beginner a speedier and richer return than thrice the quantity of upland whose salable value consists in the single fact of its being upland instead of swamp?

From my own personal experience, I can speak of the value of these swamp lands which are so thickly scattered over our country, unimproved and unappreciated, as much to the shame of their owners as to the discredit of that shrewdness which ought to be manifested by the many who are seeking ways and means to get a farm. Five years ago, I was applied to by a young man who was anxious to begin his effort to secure a home of his own. He was a capital farm-hand, possessed great energy, and had saved $200 of his earnings. After learning his views, I suggested to him the propriety of purchasing a piece of swamp land, containing twenty-six acres, which was very nearly the counterpart of that above described. It produced nothing whatever—was too wet for a cow pasture, and was in fact a neighborhood nuisance. I had long noticed it, and had studied its capabilities. I had even pointed them out to the owner, but he was one of those farmers who have little faith in any one's knowledge but their own, and he refused to believe—his only desire was to sell.

I took the young aspirant for a home all round the swamp, and into it as far as we could penetrate in a very wet season. It was grown up with alders, young dogwood bushes, and maples, with here and there a clump of tolerably large trees of other varieties. In some places where there was a slight rise in the land, it was firm and solid. I drew his attention to this circumstance, as proving that if the water could be as effectually drained away from the whole swamp as it was from these elevated spots, it must become equally dry. He recognized the reasonableness of the inference; and, after a thorough examination of the matter, assented to the feasibility of completely reclaiming the land.

But the whole condition and aspect of the swamp was so forbidding, that, although his judgment was convinced, yet he hesitated about undertaking the task. He had never drained a swamp, nor seen a similar job done by others. He spoke of the numerous waste places in the neighborhood resembling this, and of the fact that not one owner had ever undertaken the business of reclaiming a single acre, though so much wealthier than himself. He was satisfied that the redeemed land would be of the highest value, but he doubted if his means would hold out. The difficulty was to get him to begin—he had a commendable degree of courage, but not quite enough.

Finally, his hesitation was overcome by a third party offering to furnish the money with which to pay for the swamp, to wait any time for him to refund it, and, in case of his little capital and his own labor proving insufficient, to assist him with whatever more might be needed.

With this agreement to rely on the swamp was purchased at twenty dollars an acre. Taking it as it stood, this was a high price; but looking at it with reference to what could be made of it, the price was low enough. It lay within twelve miles, by rail, of a city of many thousand inhabitants, and there was a station within gunshot, from which vast quantities of milk, vegetables, and other farm products, were daily carried to the city. The railroad company was one of the few that assiduously cultivate the way traffic of the country through which they pass. The directors were as studious of the interest of the man whose whole freight was contained in a single milk-can, as of his who brought a load of wheat. Thus, let the farmer grow whatever crop he considered best, he could rely upon the railroad to convey it punctually and cheaply to the adjacent market. It is the combination of such facilities that gives value to land. Without such combination in the present case, the swamp referred to would have been dear at any price.

Hence, in the purchase and reclamation of a swamp, reference must not only be had to how cheaply it may be brought into tillage, but how near may be the market for its products, because the nearer it may be to market, the higher will be the prices to be obtained for them. The first charge on agricultural productions is that of freight—the cost of moving them to market from the spot whereon they were grown.

The swamp was a parallelogram, with a water-course running professedly from end to end, but by courses surprisingly tortuous. There was a good natural fall, but the stream had become dull and lazy from a thousand obstructions, such as fallen trees, clumps of alders, old roots, and sandbars. At its lower end an ample outlet could be created, through which any volume of water from above would flow off rapidly whenever the outlet should be opened. These important facts had been noted before the purchase was made. When the fences had been shifted by the lines of the tract, an acre of upland was found to be included, on which buildings could be placed.

My protegé began by opening an outlet at the lower end of the swamp, from which he cut a wide ditch through the whole, from one end to the other, felling trees, taking out roots, and grubbing up the dogwood and alders. Thirty days' labor of himself and one hired man, accustomed to such work, completed the ditch. The effect was immediate and very decided. Instead of the old sluggish stream, a lively current now flowed rapidly down the new water-course, into which trickled a hundred little streams from both sides of the swamp. Heretofore they had stagnated for want of an outlet—now they put on a wholesome activity. Numerous ponds and puddles quickly disappeared, while sundry powerful springs became so well defined that their sources at the foot of the upland could be distinctly identified.

While the changed condition of the land was thus enabling it to throw off a large portion of the surplus water which had made it worthless, the young owner went to work on the trees and underbrush. So great had been the change effected by the single ditch already made, that he felt greatly encouraged to persevere. As often happens, in such cases, the trees produced more cords of wood than either of us had anticipated. The brush was trimmed and bundled into faggots, which sold readily for kindling. The dogwood bushes were grubbed up bodily, so as to preserve the crook or curve at the root, and were then sold to a manufacturer in the neighboring city, who converted them into canes and umbrella handles, paying such a price as more than refunded the cost of grubbing. The wood was of course salable enough.

Five months, from April to September, were occupied in these various labors; but though a hot sun had been playing on the now exposed surface of the swamp, causing an uninterrupted evaporation of its moisture, yet in some places it was still too wet to admit of hauling off all the wood, and that portion was left until a hard freeze the following winter. Meantime the grubbing went on wherever a root was found small enough to be extracted. Many larger ones were of course left, the operation of taking them out being too expensive for a beginner. But vast piles of the smaller ones were collected and used in filling up the old tortuous water-course, while the dirt from the new ditch was wheeled over planks to cover them, leaving the roots so far under ground as to be below the reach of ordinary ploughing. This operation secured two important advantages—it cleared the new ditch of the embankments thrown up on its margin in digging it, and it brought up the old one to the surrounding level. By the first, all the surface water was allowed to flow off, as there could be no ponding or backing where no bank existed. By the last, a huge, crooked gully was converted into fast land.

But the work was not yet done. Cross drains were to be dug from the main ditch, on both its sides, extending to the adjoining upland. So far there had been no proper time to do this, as all such drains would have been choked or destroyed by the falling trees, or the operation of grubbing. But in September this work was begun on one side of the main ditch, and was pushed rapidly to completion. Narrow drains or ditches were sunk wherever they were supposed to be necessary, in the bottom of which small bushes were deposited and then covered. Care was taken to lay the bushes all one way—lengthwise. Each one of these underdrains began immediately to perform its office of relieving the ground of a portion of its surplus water, as might be seen by examining the outlet in the main ditch. Some of them, such as had tapped a spring, discharged copious streams.

All who have had any experience in underdraining are aware how quickly it transforms wet land into dry land; and those who have not, can form no idea of it until they witness the result for themselves. It was so with my protegé. He was astonished at the work of his own hands. The drained side of the swamp now dried up very rapidly. The land in some places settled away from the large stumps which yet remained, and was evidently becoming hard and firm. These indications kept him in excellent spirits, to which the manure-like appearance of the rich black soil that had everywhere been turned up as the drains were dug, made a further contribution. No one could walk over the ground on the two sides of the main ditch without instantly discovering the difference between the drained and the undrained. That winter he employed in clearing up the swamp more thoroughly, being now of opinion that he might be able to plough it in the spring.

When spring came, he did succeed in ploughing nearly all the underdrained half, then limed it, planted corn, and raised the first crop that that swamp had ever produced. The corn was so large as to be an amazement to the neighborhood. It exceeded the yield of the upland on both sides of his line, and settled the question as to the profitableness of reclaiming swamp lands.

The young owner was so much encouraged by his success that, after getting in his corn, he immediately proceeded to underdrain the remaining half. The following spring it was dry enough to be all ploughed. He then limed it, and at the proper season sowed buckwheat, securing a crop quite as heavy as any of his neighbors. These first crops were regularly succeeded by others, and they were invariably good ones. In time, as the vegetable mould decayed, the soil became loose and of remarkably easy cultivation, while in richness it far exceeded that of the best upland in the neighborhood. It needed no manure. The more it was turned up to the sun, the drier and more friable it became. Fruit trees were planted and flourished, and strawberries were grown in the rich new soil with astonishing success. It seemed to be the very home for cabbages, turnips, and celery; and there is now as little prospect of its relapsing into swamp land as there was, ten years ago, of its graduating into arable land. It could be sold any day for $120 an acre.

So much for its marketable value, but now for the

cost of reclaiming it. This, all told, amounted to $21 an acre, but not including the young owner's time. I make no account of that, because he has his pay in the increased value of the land. The actual money-cost was more than $21 an acre, but the total stands at that figure by deducting sales of wood and faggots.

There are tracts of swamp land which contain three times the quantity of timber that this did, which can be purchased quite as cheaply, and which, from the greater quantity of wood they might yield, could be reclaimed to better profit. There are others, wholly clear of wood and underbrush, which could be reclaimed for even less. It should be the business of the shrewd and enterprising to seek out and appropriate them.

The first cost of this swamp was $20 per acre, the reclaiming of it $21, making a total of $41, or $1066 for the whole. Its value rose in two years to $3120, or very nearly treble the first outlay. Here was a large capital suddenly created out of a small one, not by mere investment of the original sum, but by bringing to its aid the experience of one man and the courageous industry of another. It was the judicious combination of the three upon a specific object, that worked the change. Thus, the man who improves his own land, works with a long lever, and will find that, in reality, but little power is required. The lesson should be as instructive to those who read this, as was the reclaiming of the swamp referred to to the neighborhood wherein it lay. The owner's success was so decided, that it sent up the price of numerous waste places of similar character, and caused others to be as effectually drained. Exhortation had been lost upon the owners; but when the example was placed before them, imitation came of itself.

The young man whose personal energy converted twenty-six acres of utterly waste land into a productive farm, created for himself, at a single stroke, a capital that set him up for life. He proved conclusively that it is the best lands which come last into cultivation. The single acre of upland in his purchase now contains his house and barns, and from the remaining acres he produces all that his family needs. In ten years from the day that he first struck his spade into the main ditch of an apparently worthless swamp, he will be out of debt, and worth his thousands.

In the estimation of some, an undertaking of this character will smack of speculation. It is out of the old routine—it is a new way to get a farm—and being new, is therefore speculative, and being speculative, is not only foolish, but hazardous. But here the prevailing ingredients are good sense and resolute industry. The speculators hate work—the industrious hate speculation. Timid minds will reject the example just given, not only because constitutionally fearful of a new thing, but because of an equally constitutional incredulity. But the experience of many in this country could be adduced in confirmation of the idea that one of the surest ways to get a farm cheaply, is to purchase and reclaim a swamp. It involves hard work as well as dirty work, wet feet and muddy clothes, but few undertakings will pay better.

It is known that several counties in New Jersey contain thousands of acres of cranberry lands, which annually produce abundant crops of fruit. In numerous locations the owners of the land receive no part of the crop. The whole region is but thinly settled, and there are but few clearings among the dense pine forests which cover a large portion of the ground. Most of these have been made by pine-hawkers and charcoal-burners, who support life under great privations, and who rear families in total ignorance of schools or churches. All round them lay immense cranberry grounds, without a panel of fencing on thousands of acres. From time immemorial these squalid families have gathered the fruit for their own benefit, and disposed of it at the nearest stores. They swarm among the swamps during the picking season, so that the owner gets little or none of the crop. If residing at a distance, as is generally the case, he has no chance whatever. Even when within a few miles of his own swamp, he receives no portion except by sufferance. The cranberry grounds of the region have been so long abandoned to these indiscriminate inroads, that the annual plunder of the crop has grown to be considered a public right. In some places, the owner may receive a barrel or two as a gift, in others he is permitted to send in a few pickers for his own use. But it would be a dangerous experiment for him to undertake suddenly and forcibly to suppress these depredations—a general mutiny would be the result.

All these wild lands are consequently unsalable, and are worth from two to five dollars per acre. They have been in market at these prices for many years. Some of them are covered with small pines and scrub oak, the latter growing up wherever the former have been cut off for the use of iron furnaces or glass works. Large districts of them are within a few miles of a railroad, on which two hours' travel will convey you to Philadelphia, and three more to New York. But they lie on no public thoroughfare, and hence they have remained unnoticed and neglected. No good roads penetrate them, hence there are no travellers, and hence the distant public knows but little of their capabilities.

Seven years ago an enterprising farmer of Burlington county purchased a hundred acres of this swamp land, for which he paid $5 per acre. All the adjoining land was in market at the same price, a third or a half to be paid down. Much of his purchase was grown up with cranberries. His design was to give them some general attention that would cost but little money, and if found to pay, then to bestow more care upon them, and eventually to convert the whole tract into a carefully cultivated cranberry plantation. He conciliated, to some extent, the jealousies of the neighboring sand-hillers by employing them to cut brush sufficient to construct a fence around the portion to be protected, so as to keep off the hogs and cattle which roamed the woods, rooting up or trampling down the plants.

This slight protection of a brush fence, costing but little, produced the best results. The plants bore freely, and ripened their fruit well. He had purchased the land in April, and that fall he sold $600 worth of fruit, netting $440. The following year was less productive, in consequence of a heavy frost when the plants were in bloom. But the third year paid better than the first. In the mean time he had enlarged his plantation by extending his brush fence around other portions of the tract, and at this writing is clearing an average of $1100 annually from about thirty acres.

In the sections of New Jersey referred to, there are numberless places whereon similar operations may be carried out. Some of them have been already redeemed from their native wildness by a very moderate expenditure, and converted into the most lucrative investments. Once established, these cranberry swamps rarely fail. Now and then a frost may injure the crop, or the worm may wholly destroy it; but on the average of five years, there is probably no investment that can be made to pay better. A cranberry swamp, well set with vines, and conveniently located, is a cheaper purchase at $50 per acre than a quarter section of government land as a gift. If so located as to be easily flooded, the crop may be considered sure.

For a beginner it possesses rare advantages. Generally it will take care of itself, requiring little labor or attention, except when the crop comes in. For at least ten months of the year he may employ most of his time in working out for others, or in cultivating other land for the production of food for his family. The crop is among the most marketable of all the fruits. Its sale is distributed over the whole year, instead of being, like other berries, crowded into a few weeks. It is largely exported, and there has rarely been a glut.

At a meeting of the New York Farmers' Club, in 1858, a letter from Mr. Edwin Salter, of Barnegat, Ocean County, New Jersey, was read, from which the following extract is taken:

"I notice, at a late meeting of the Farmers' Club, that the subject of transplanting cranberries was brought up. Could you put me in a way to find out what the New England cranberry-growers consider a good crop, natural growth and transplanted? There are scores of acres in this vicinity that yield often over 100 bushels per acre from natural growth; transplanting is here a new thing, and has not been carried on long enough to know how successful it will prove. The heaviest yielding cranberry-bog in this section is one in the woods some twenty miles from any habitation (except cabins), which is said to have yielded 300 bushels to the acre one season. In the vicinity of this bog are hundreds of acres of land which appears to be naturally such as at your meeting was described as the best adapted for transplanting. As this land is held at only a nominal price (from $1 to $3 per acre), it would doubtless pay to try transplanting there. At any rate it will be tried. It ought to pay better than New England land, for which, in addition to the higher price of land, heavy expenses sometimes have to be incurred to make it precisely what this is naturally. Our shore people here are nearly all seafaring men, but of late some few of them are turning their attention to the soil; their experiments thus far prove that, though our land may not be as good as in other sections, yet it will prove as profitable to till, especially as we are but a day's journey from both New York and Philadelphia markets."

If it be granted that it is a good thing to get a farm, it would seem to be a better one to find it already planted with a permanent crop. Such is the condition of a natural cranberry swamp. But if the plants there flourish with profitable luxuriance, it does not follow that they can be cultivated with advantage on no other description of land. There are varieties which have been proved to be successful even on dry upland. But as such land is costly, its use for this culture does not properly belong to the subject in hand. Yet on the pine barrens of Long Island, these berries have been raised at the rate of seventy-five bushels per acre, after being set three years, the sod from the native swamp, full of plants, being transferred directly to the upland. In various parts of New England, the upland culture is reported to be profitable. But for light on this question the inquirer is referred to such books as treat fully of the cranberry culture.

The reader of any of the numerous agricultural periodicals issued among us, cannot fail to have noticed repeated narratives of success in the reclamation of swamp lands. There are many instances in which the work has been accomplished at a surprisingly low cost, while in others it has been the reverse. It should be the study of the beginner to so select the spot on which he is to operate, as to take that only which can be reclaimed at the minimum expense. Our country abounds with desirable fields for such operations. Every reader can call to mind some worthless swamp or cripple, admirably situated as to neighborhood and market, large enough to make a respectable farm, and obtainable at a low price on such terms of payment as a small capitalist could comply with. In some locations the neighbors would be disposed to aid the man who had enterprise enough to undertake the abatement of a nuisance which was not only offensive to the eye, but dangerous to public health.

The surpassing richness of such lands, when redeemed from the dominion of the water, has long been proverbial. They need no manure, yet they produce double crops. Strangely enough, though the best soil on the continent, yet it comes last into cultivation. Hence I am free to urge this process as one of the best, the quickest, and the surest, by which to get a farm. Like the cranberry swamp, it possesses peculiar advantages for a small beginner, as it yields immediate returns, and as there are long intervals between the different stages of work, during which he can employ himself at other labor. When his drains have once been laid, the land will be drying while he sleeps; and if, when seeking for a farm, he had been recommended to plunge into a swamp, he will find, in the end, that it was not asking for bread and receiving only a stone.

Such an example operates powerfully on others. Mr. Beecher says, that "if men are to become intelligent, we must give them specimens of intelligence. Let a man go into a village where the houses are all going to decay, where the fences are all tumbling down, and where no pains are taken with trees and flowers, and build a neat house, inclosing his grounds with a neat fence, and tastefully decorating his yard with comely trees and beautiful flowers, and his example will be a blessing to the place. In three years there will be twenty neat houses, with good fences, and yards blooming with shrubbery and flowers, as the result of his judicious outlay of means. The taste of the whole village will be educated and improved by the influence he will exert through the instrumentality of whatever advantages he may possess over its inhabitants."

But if moderate capitals may be advantageously invested in the reclaiming of either swamp or cranberry lands, it does not follow that large ones, so used, will fail to produce equally satisfactory returns. One instance may be cited where a large sum was devoted to reclaiming a cranberry swamp in Massachusetts. The leading particulars were given in the Boston *Cultivator*, from which the following account is taken. The grounds were visited by the editor in November, 1863. Their owner is Dr. E. D. Miller, whose residence is Dorchester, some twenty-five miles from Boston. The swamp is described as having been almost worthless.

"Something like ten years since, this swamp was covered over with a growth of alders, dogwood, white maples, and other swamp-shrubs, which covered the ground; they were cleared off, and a ditch cut through the swamp for the brook, which before ran through a very crooked channel. Ditches were then opened from the uplands on each side, which are gravelly and sandy, leading into the main ditch. A dam was constructed across the swamp, which serves the purpose of flowing it, and also that of a road to pass across it. In the winter, the swamp was usually flowed, and gravel, this being better than sand, was drawn on the ice and spread. Afterwards it was planted to cranberry cuttings, in drills about eighteen inches apart, this, from experience, proving to be a suitable distance apart. How many coverings of gravel have been put on was not learned; but several, judging from the excavations when removed.

"About twelve or fourteen acres of this swamp have been planted; and so favorably is it situated, that it can be covered with water in little more than an hour's time. The brook is of such capacity, with the aid of a reservoir above the cultivated ground, that the plants can be protected from frosts at any season when there is any danger.

"The crop of the past season was about 1100 barrels of very nice fruit, and of remarkable size. I brought away a couple of berries, that measured nearly three inches in circumference. The crop was all picked by hand, at a cost of nearly $2000. At one time, said Dr. Miller's farmer, 200 persons might have been seen in that swamp picking cranberries. It was a lively scene. After they were gathered, they were taken to the house, where they were sorted, the soft berries, after winnowing them, were culled out by women and girls, preparatory to barrelling.

"When Dr. Miller first contemplated the cranberry culture of this swamp, he visited Mr. Joseph Breck, the well-known seedsman of North Market-street, Boston, and asked him how to go to work. Mr. Breck said he could not tell him: then he asked for the best work on cranberry culture. Mr. B. told him he did not know of any he could recommend; then, said Dr. Miller, 'Can you tell me of a man I can employ that knows something about it?'

and Mr. Breck said he could not. 'Well,' replied Dr. M., 'then I will try and see what I can do.' The result and the mode of doing it is briefly stated as above, as learned from Dr. Miller and Mr. Desmond, his farmer.

"Dr. M. has informed the writer, since visiting the cranberry swamp, that the fruit has generally been sold so far as it is marketed, at the current price, though some of it was sold at $15 a barrel. Call the average price $10 a barrel, and 1100 barrels will bring the snug little sum of $11,000. This beats tobacco-raising out of sight, as the saying is.

"One of the peculiar advantages possessed by Dr. Miller over most of the owners of swamp lands, is the facility with which he can flow it at all seasons of the year, thus guarding the growing crop from both late spring frosts and early autumn frosts; and, besides, gives him the power to destroy insects that sometimes infest the vines. Swamp lands that can be as quickly flowed and as quickly drained as Dr. Miller's cannot be used more profitably than by growing cranberries, as it would seem by the Doctor's experience. It is also easily gravelled in the winter by flowing it."

The foregoing account was considered so remarkable, that I applied to Dr. Miller for some additional items of information. From these, in connection with the narrative quoted, the reader will be able to form a just conception of the magnitude of the enterprise, of the amount of capital invested, as well as of the character of the result.

Dr. Miller has about twenty-five acres, divided into five meadows, varying in size from 3 to 12 or 14 acres, all on the same stream of water. The whole can be overflowed at will in about two hours.

About 18 to 20 acres are drained by side ditches discharging into a main ditch, all being open—no underdrains. The top soil of the meadow was removed to the depth of eight to eighteen inches, and about six inches of coarse gravel was put in its place. The cranberry plants which were growing wild upon the spot, were necessarily removed by this operation, but they were replanted on the gravel. The principal meadow of 12 to 14 acres was purchased at $12 per acre, and the cost of reclaiming and planting was about $500 per acre.

The work was begun in 1852, and the next year a small quantity of fruit was gathered. The succeeding crops, for eight years, varied from 200 down to 7 barrels. The crop of 1862 was the smallest of all, and was lost by the neglect of a workman; that of 1863 was the highest, and amounted to 1030 barrels. Dr. Miller says, "I have no doubt that when my meadows are brought to a proper condition, they will yield, some seasons, from 2000 to 2500 barrels."

He says, moreover, that his example is followed by others to a limited extent, owing to the want of suitable meadows as to flowing, gravel, &c. He thinks he could now do the same amount of work (with labor at the same price) for $150 less per acre. It was to get rid of the multitude of roots in the ground that he removed so great a depth of soil. All this was taken away by wheelbarrows, the ground being too soft to sustain the tread of cattle. It was finally burned, and the ashes used on other land. The larger meadow had run to waste so long, that for some time the owner could not decide to whom it belonged.

It will doubtless occur to the intelligent reader that Dr. Miller committed an oversight in destroying by fire the immense amount of sods, grass-roots, muck, and peat, which he thus removed with the alder and other roots. From so large a surface, and going to so great a depth, the accumulation must have been enormous. Though when dry it would readily burn, yet it was in reality a highly concentrated manure. There was a certain value in every load, as would have been manifested had it been applied to other land. But the Doctor admits that he kept a loose account of the profits he gained by removing this mass of vegetable matter, and says, "I was more anxious to get rid of it than to make anything out of it." As he managed the affair, he thinks he realized about 1200 bushels of ashes. But it may be safely assumed that the real value of the manure thus passed through the destructive ordeal of fire, was fully equal to a third, if not the half, of the whole cost of removing and replacing it.

This estimate is warranted by my own experience. Some four years since I began the operation of reclaiming a jungle of three acres, which, since the foundation of the world, had produced no crop but alders, briars, hassocks, ferns, and bumble-bees. It was too soft for cattle to graze through it, and in many places would almost mire a goose. The alders had been cut off thirty years previous, but the roots had been left in the ground. For all agricultural purposes this meadow was absolutely worthless.

I began by opening a ditch through its whole length, some 800 feet, then dug narrow cross-ditches at right angles, 30 feet asunder, in which were deposited trunks formed by nailing two boards together, and discharging into the main ditch. They were then covered, and became underdrains. This was all done in the autumn. In July following, the ground was dry and hard enough for a team of four mules to plow up the tough and almost impenetrable sod that covered it. In October this sod was taken off and collected in a huge pile, so high and so long as to resemble a railroad embankment. Two hundred bushels of lime were mixed in as the pile was made. That winter the denuded meadow was filled in with dirt from an adjoining highland, in many places to the depth of a foot, the wheelbarrow and horse-shovel being used. The next season it was all ploughed and planted with corn, cabbages, and pumpkins.

But the sod taken from the surface was worth more than the whole improvement cost. It amounted to several thousand loads. It burned readily when only half dried, and could have been as quickly converted into ashes as that which fell to the lot of Dr. Miller. But in place of being burned, a section of the heap went four times a year to the barn-yard, where it speedily graduated into the richest kind of manure.

It was interesting to notice how every plant grew and flourished to which it had been freely applied. Corn, grapes, strawberries, celery, potatoes, and garden vegetables especially, were stimulated into a most luxuriant growth. It could hardly be otherwise, as it was pure vegetable fibre in a state of decay. Some professional florists who visited my premises, and examined this great muck-heap, expressed their admiration of its richness. It was the very material to obtain which they were compelled to send many miles, and without which it would be scarcely possible for them to prosecute their business. I could have readily sold it for more money than the whole improvement of the three acres of meadow cost me. This experience is cited as one of the forms in which the reclaiming of a swamp will return immediate compensation.

The abundance of swamp lands in this country having been referred to, it may be of some interest to give such figures in relation to the subject as may be relied on. There are no statistics showing how many acres the whole Union contains; but some idea may be gathered from the legislation which has been had in Congress when coming to dispose of them. The first statute which dealt with this interest was the law of 1849, in relation to Louisiana, a large extent of whose territory was annually overflowed. Along the Mississippi the alluvial margin is from one to two miles wide; and to prevent the inundation from that river, an artificial embankment or levèe system had been adopted, extending, on the east side of the river, from forty miles below New Orleans to a distance up the river of a hundred and eighty miles, and on the west side generally to the boundary of Arkansas.

To aid Louisiana in constructing the necessary

levèes and drains to reclaim these swamps and overflowed lands, Congress, by act of March 3, 1849, granted to that State "the whole of those swamps and overflowed lands which may be, or are found unfit for cultivation." The Government, in the spirit of enlarged public policy, conceded this class of inundated lands to aid in the construction of permanent levèes, with a view to secure private property, and also as a sanitary measure. To this succeeded the law of 1850, extending a similar grant to Arkansas; but the last section of the act enlarged the grant, so as to embrace "each of the other States of the Union in which such swamp and overflowed lands may be situated." When this measure had its origin, and before it became general, the grant was estimated as embracing 5,000,000 acres. But in September, 1863, it was ascertained that the enormous quantity of 57,923,737 acres had been selected.

How such extraordinary flexibility was imparted to a law having a definite object, is shown by the subjoined explanation from the *Washington Union:*

"It is currently reported that extensive frauds have been attempted in regard to the selection of swamp lands under the act of September 28, 1850. That act granted to the States the swamp and overflowed lands, unfit thereby for cultivation, which were at that time unsold. Some of the States selected swamp lands in accordance with the field notes of the Surveyor-General; other States appointed agents to select these lands, the agents furnishing lists to the General Land-office, which lists, having been examined by the Surveyor-General, were reported for approval or disapproval. In some of the States these lands were granted to the counties in which they were found, by the State Legislatures. The counties, in some instances, entered into contracts with the agents for the purpose of selecting those lands for them. It is said that, in some instances, these agents went into the fields and selected all the good vacant land which they could find, irrespective of its character, whether swamp or otherwise. These agents, by contract, were allowed, say ten or fifteen cents an acre in some instances, and in others one-fourth and one-third of the lands found. Under these tempting inducements, some swamp lands have been found on the tops of high hills and mountains. If these lists of selections by these agents should have been sanctioned by the Department, this class of speculators would have made from $250 to $50,000 a day. On the 3d of March, 1857, an act was passed in relation to these selections, which the Department of the Interior holds does not relate to selections of lands made after the date of the act itself. It seemed to be the design of this act of 3d of March, 1857, to confirm to the several States such lands as may have been selected under the act of 28th September, 1850, which had heretofore been reported to the Commissioner of the General Land-office, so far as then vacant and unappropriated, and not interfered with by actual settlement. Selections which have been made since the date of this act, it will be perceived, have not been confirmed thereby. Some of these parties, who have expected to be benefited by this act of 1850, will find themselves sadly disappointed."

In the overflowed region along the Mississippi river, for the reclamation of which the law was originally intended, and where it went earliest into operation, the effect was speedy and very remarkable. During the three years ending with 1857, there was an increase of over $50,000,000 in the assessed value of the lands in Mississippi, and of over $25,000,000 in the value of the taxable slaves.

A very large portion of this enormous increase of value occurred in the regions subject to overflow from the river, but which, under the operation of the law just referred to, had been drained and brought under tillage. But this system of drainage had scarcely been begun. All thus far accomplished had been hastily, and therefore imperfectly done. Yet the amazing fertility of the half-reclaimed bottoms is shown in the rapid increase of values. Their superior productiveness attracted capital and population from all the slave States. Within the three years mentioned, the slave population of Mississippi increased from 326,861 in 1854, to 368,861 in 1857. The progress developed in Arkansas and Texas was even more remarkable. This increase would have gone on enlarging as the drainage extended and became more complete. But rebellion not only interrupted its progress, but made the whole region a hissing and an astonishment to the world. As an organized institution, slavery became utterly demoralized, and in many places ceased to exist. The slaves became fugitives or soldiers. Thousands perished of disease or starvation. Plantations were abandoned, their buildings and machinery destroyed, leaving what is probably the finest region on the continent to be repeopled and resuscitated by a new race of owners.

This appropriation of public land may be regarded as the only instance in which our Government has extended aid to promote the drainage of any description of lands. While with us the neglect of this important interest has been the rule, in England the practice has been directly the reverse. For a long series of years, the efforts of English landlords have been directed to the reclamation of waste lands, moors, heaths, and lowlands. In 1797, a committee of the House of Commons estimated the area of such lands as had been brought under inclosure from 1700 to that date, at about 4,000,000 acres. The subsequent increase was relatively much greater, as there are statistics to show that from 1800 to 1820, as many as 3,000,000 acres more were brought under tillage. It is held that this rapid addition was caused by the stimulant of high prices for all agricultural products growing out of the wars of that period.

Since 1820, it is said that the reclamation of waste lands in England has not been pushed so vigorously, but that the effort has been to increase the acreable product of land, rather than to enlarge the area. Hence the enthusiasm touching artificial fertilizers and underdraining, both looking to a larger expenditure of capital and labor on an acre. Government has shared in this enthusiasm by loaning to English farmers some $60,000,000 to enable them to underdrain their lands. It is among the remarkable facts of this munificent loan, that the lender has sustained no loss from the borrowers, and that no land has been underdrained without being signally benefited. In numberless instances its productive capacity has been doubled. Indeed, some authorities have held, that but for the increased supply of food thus produced in England, as the result of a vast system of underdraining, the people of that country would be almost as helplessly dependent on other nations for food, as their manufacturing industry has been dependent upon us for cotton.

Thus, if underdraining has been recognized in England as a subject deserving of national encouragement, the fact is a significant endorsement of the prominent position which it has been made to occupy in this volume. But though no similar aid has been given to American farmers, yet there are thousands of them who, aware of its importance, have made drainage a fundamental element of their whole system of farming. From among these a single instance may be cited, as showing not only how thoroughly the art has been transplanted to this country, but how fully its results here corroborate those which have been realized by English farmers. The facts are taken from the New York *Tribune*, for October 29, 1859:

"Mr. John Johnston, near Geneva, N. Y., at one time esteemed a fanatic by his neighbors, has come, of late years, to be generally known as 'the father of tile-drainage in America.' After thirty years of precept, and twenty-two of example, he has the satisfaction of seeing his favorite theory fully accepted, and to some extent practically applied throughout the country. Not without labor, however, nor without much skepticism, ridicule, and controversy, has this end been attained; and if, now that his head is whitened, and his course all but run, he finds himself respected and appealed to by persons in every State of the Union, he does not forget that it has been through much tribulation that he has worked out this exceeding great weight of glory. Mr. Johnston is a Scotchman, who came to this country thirty-nine years ago, and purchased the farm he now occupies on the easterly shore of Seneca Lake, a short distance from Geneva. With the pertinacity of his nation, he stayed where he first settled, through ill fortune and prosperity, wisely concluding that by always bettering his farm he would better himself, and make more money in the long run than he could by shifting uneasily from place to place in search of sudden wealth. He was poor enough at the commencement; but what did that matter to a frugal, industrious man, willing to live within his means, and work hard to increase them? And so with unflagging zeal he has gone on from that day to this.

"His first purchase was 112 acres of land, well situated, but said to be the poorest in the county. He knew better than that, however, for although the previous tenant had all but starved upon it, and the neighbors told him such would be his own fate, he had seen poorer land forced to yield large crops in the old country, and so he concluded to try the chances for life or death. The soil was a heavy gravelly clay, with a tenacious clay subsoil, a perfectly tight reservoir for water, cold, hard-baked, and cropped down to about the last gasp. The magician commenced his work. He found in the barn-yard a great pile of manure, the accumulations of years, well rotted, black as ink, and 'as mellow as an ash-heap.' This he put on as much land as possible, at the rate of *seventy-five loads to the acre*, ploughed it in deeply, sowed his grain, cleaned out the weeds as well as he could, and the land on which he was to starve gave him about twenty-five bushels of wheat per acre. The result was, as usual, attributed to luck, and any thing but the real cause. To turn over such deep furrows was sheer folly, and such heavy dressings of manure would not fail to destroy the seed. But it didn't; and let our farmers remember that it never will; and if they wish to get rich, let them cut out this article, read it often, and follow the example of our fanatical Scotch friend.

"This system of deep ploughing and heavy manuring wrought its results in due time. Paying off his debt, putting up buildings, and purchasing stock each year to fatten and sell, Mr. Johnston, after seventeen years of hard work, at last found himself ready to incur a new debt, and to commence laying tile drains. Of the benefits to be derived from drainage he had long been aware; for he recollected that when he was only ten years of age, his grandfather, a thrifty farmer in the Lothians, seeing the good effects of some stone drains laid down upon his place, had said, 'Varily, I believe the whole airth should be drained.' This quaint saying, which needs but little qualification, made a lasting impression on the mind of the boy, that was to be tested by the man, to the permanent benefit of this country.

"Without sufficient means himself, he applied for a loan to the bank in Geneva, and the President, knowing his integrity and industry, granted his request. In 1835, tiles were not made in this country, so Mr. Johnston imported some as samples, and a quantity of the 'horse-shoe' pattern were made in 1838, at Waterloo. There was no machine for producing them, so they were made by hand, and molded over a stick. This slow and laborious process brought their cost to $24 per thousand, but even at this enormous price Mr. Johnston determined to use them. His ditches were opened and his tile laid, and then what sport for the neighbors! They poked fun at the deluded man; they came and counseled with him, all the while watching his bright eye and intelligent face for signs of lunacy; they went by wagging their heads, and saying 'Aha!' and one and all said he was a most consummate ass to put crockery under ground and bury his money so fruitlessly. Poor Mr. Johnston! he says he really felt ashamed of himself for trying the new plan, and when people riding past the house would shout at him, and make contemptuous signs, he was sore-hearted, and almost ready to conceal his crime. BUT WHAT WAS THE RESULT? Why this: that land which previously was sodden with water and utterly unfruitful, in one season was covered with luxuriant crops, and the jeering skeptics were utterly confounded; that in two crops all his outlay for tiles and labor was repaid, and he could start afresh and drain more land; that the profit was so manifest as to induce him to extend his operations each succeeding year, and so go on until 1856, when his labor was finished, after having laid 210,000 tiles, or more than fifty miles in length! And the fame of this individual success going forth, one and another duplicated his experiment, and were rewarded according to their deserts.

"It was not long after the manufacture of the first lot of tiles that a machine was contrived which would make them quite as well and faster; and by its aid they were afforded at quite as low a price as after an English machine was imported. The horse-shoe tile has been used by Mr. Johnston almost exclusively, for the reason that they were the only kind to be procured at first, and on his hard subsoil, finding them to do as well as he could wish, he has not cared to make new experiments. He has drains that have been in function for more than twenty years without needing repair, and are apparently as efficient now as they were when first laid. In soft land, pipe or sole tiles would be preferable, or if horse-shoe were used, they should be placed on strips of rough board, to prevent them sinking into the trench bottom, or being thrown out of the regular fall by being undermined by the running water. He has not used the plough for opening his trenches, for the reason that all his work has been let out by contract, and the men have opened them by the spade; charging from twelve and a-half to fifteen cents per rod for opening and making the bottom ready for the tile. The laying and filling was done by the owner.

"His ditches are dug only two and a-half feet deep, and thirteen inches wide at the top, sloping inward to the bottom, where they are just wide enough to take the tile. One main drain, in which are placed two four-inch tiles set eight inches apart, with an arch-piece of tile having a nine-inch span set on top of them, was dug three and a-half and four feet deep, and this serves as a conduit for the water from a large system of laterals. Drains should never be left open in winter, for the dirt dislodged by frequent frosts so fills the bottom that it will cost five or six cents per rod to clear them; and, moreover, the banks often become so crumbled away that the ditch cannot be straddled by a team of horses, and thus most of the filling must be done by hand. Mr. Johnston, in draining a field, commences at the foot of each ditch, and works up to the head. He opens his mains first, and then the lateral or small drains, but he lays the tiles in the laterals, and fills them completely, before laying the pipe in the mains. The object of this is to prevent the accumulation of sediment in the mains which would naturally be washed from the laterals on their first being laid. By commencing at the foot of each ditch, and working upward, he can always get and preserve the regular fall, which may be dictated by the features of his field, more easily than by working toward the outlet. A little practice teaches the ditchers how to preserve the grade almost as well as if gauges were employed; but before laying the tiles, the instrument is applied to test the bottoms thoroughly. The necessity of this precaution will be apparent to any one who reflects that if a tile or two in the course of a ditch be set much too high or too low at either end, the water quickly forms a basin beneath and around, sediment is washed into the adjoining pipe, and ultimately even the whole bore is filled and the drain stopped. When this happens, it will be indicated after a time by the water appearing at the surface of the ground above the spot—drawn upward by capillary attraction. In such a case the ditch must be reopened and the tile relaid.

"Mr. Johnston says tile-draining pays for itself in two seasons, sometimes in one. Thus, in 1847, he bought a piece of ten acres to get an outlet for his drains. It was a perfect quagmire, covered with coarse aquatic grasses, and so unfruitful that it would not give back the seed sown upon it. In 1848, a crop of corn was taken from it, which was measured, and found to be *eighty bushels* per acre, and as, because of the Irish famine, corn was worth $1 per bushel that year, this crop paid not only all the expense of drainage, but the first cost of the land as well.

"Another piece of twenty acres, adjoining the farm of the late John Delafield, was wet, and would never bring more than ten bushels of corn per acre. This was drained at a great cost, nearly $30 per acre. The first crop after this was 83 bushels and some odd pounds per acre. It was weighed and measured by Mr. Delafield, and the County Society awarded a premium to Mr. Johnston. Eight acres and some rods of this land, at one side, averaged 94 bushels, or the trifling increase of 84 bushels per acre over what it would bear before those insignificant clay tiles were buried in the ground. But this increase of crop is not the only profit of drainage; for Mr. Johnston says that on drained land one-half the usual quantity of manure suffices to give maximum crops. It is not difficult to find a reason for this. When the soil is sodden with water, air cannot enter to any extent, and hence oxygen cannot eat off the surfaces of soil-particles and prepare food for plants; thus the plant must in great measure depend on the manure for sustenance, and of course the more this is the case, the more manure must be applied to get good crops. This is one reason, but there are others which we might adduce if one good one were not sufficient.

"Mr. Johnston says he never made money until he drained, and so convinced is he of the benefits accruing from the practice, that he would not hesitate—as he did not when the result was much more uncertain than at present—to borrow money to drain. Drains well laid endure, but unless a farmer intends doing the job well, he had best leave it alone, and grow poor, and move out West, and all that sort of thing. Occupiers of apparently dry land are not safe in concluding that they need not go to the expense of draining, for if they will but dig a three-foot ditch in even the dryest soil, water will be found in the bottom at the end of eight hours, and if it does come, then draining will pay for itself speedily. For instance, Mr. Johnson had a lot of thirteen acres on the shore of the lake, where the bank at the foot of the lot was perpendicular to the depth of thirty or forty feet. He supposed from this fact, and because the surface seemed very dry, that he had no need to drain it. But somehow he lost his crops continually, and as he had put them in as well as he knew how, he naturally concluded that he must lay some tile. So he engaged an Irishman to open a ditch, with a proviso, that if water should come into it in eight hours, he would drain the entire piece. The top soil was so hard and dry as to need an application of the pick, but at the depth of a foot it was found to be so wet and soft that a spade could easily be sunk to the entire depth of the handle. The ditches were made, and in less than the specified time, a brave lot of water flowed in. The piece was thoroughly drained, and the result was an immense crop of corn. The field has regularly borne 60 to 70 bushels since. Corn was planted for a first crop in this and the preceding instances, because a paying crop is obtained in one year, whereas, if wheat were sown, it would be necessary to wait two seasons. He always drains when the field is in grass, if possible, for the ditches can be made more easily; and spring is chosen that the labor may not be interfered with by frosts.

"To show how necessary it is to avoid planting trees over drains, we quote a case in point. In a lot adjoining his house are four large elms, which are marked to be felled, and for the reason that the lot was formerly so wet that a pond of water stood upon it in winter, and throughout the season the children skated and slid upon it. It was drained, and all went well for a time; but after three years Mr. Johnston found that his drains did not discharge properly, and that in certain places the water came to the surface, so as to destroy or greatly lessen the crop above them. He could not account for the circumstance until he dug down to the drain at each of these spots, when, to his surprise, he found the tile completely choked with fibrous roots of the elms, which, naturally seeking the subterranean supply of water, had so accumulated in mass as to stop a two-inch bore of tile.

"Mr. Johnston does not think there are a hundred acres in any neighborhood that do not need draining, and would not pay well for it. Perhaps this may be thought an extreme assertion, but it is nearer the fact than most of us have been aware. Mr. Johnston is no rich man who has carried a favorite hobby without regard to cost or profit. He is a hard-working Scotch farmer who commenced a poor man, borrowed money to drain his land, has gradually extended his operations, and is now reaping the benefits, in having crops of forty bushels of wheat to the acre. He is a gray-haired Nestor, who, after accumulating the experience of a long life, is now, at seventy-five years of age, written to by strangers in every State of the Union for information, not only in drainage matters, but all cognate branches of farming. He sits in his homestead a veritable Humboldt in his way, dispensing information cheerfully through our agricultural papers, and to private correspondents, of whom he has recorded 164 who applied to him last year. His opinions are, therefore, worth more than those of a host of theoretical men, who write without practice. He says that the retrogression of our agriculture in the older States is to be accounted for in our lack of drainage, poor feeding of stock, which results in giving a small quantity of poor manure, and in not keeping enough to make manure. He applies 100 loads of manure to the acre at the beginning of a rotation, and this lasts throughout the course. He learned from his grandfather that no farmer could afford to keep any animal that did not improve on his hands, and that as soon as it was in good marketable condition it should be sold, and replaced by another. This theory he has always carried out, and, as a natural consequence, has always got higher prices for his beef stock, and a ready market even in the dullest times.

"Although his farm is mainly devoted to wheat, yet a considerable area of meadow and some pasture has been retained. He now owns about 300 acres of land. The yield of wheat has been 40 bushels this year, and in former seasons, when his neighbors were reaping 8, 10, or 15 bushels, he has had 30 and 40. We are informed by him that there has been no such crop as the present since 1845, either in yield or quality; and the absence of weevil is remarkable. A variety of white wheat from Missouri, sown more thinly than usual, has yielded 31 bushels to something less than one bushel of seed sown. It headed out a fortnight earlier than the Soule's, but ripened later—probably because thinly sown. Mr. Johnston thinks we have been sowing too thickly for fifteen years past upon rich land, and there can be no question but that he is right. Still, it is better to take a medium course between thick and thin sowing, and thus avoid, on the one hand, rust, overcrowding, and waste of seed, and on the other, placing an entire crop at the mercy of insects which may attack it.

"A too common error with improving farmers is that of using too small tile for main drains, and too large for laterals. Those accustomed to the roomy conduits of ordinary stone drains, suppose that nothing less than a three-inch bore will conduct the drainage from the surface into the mains, and curiously enough the same persons, unmindful of the large area drained by each system of laterals, err in using mains but little larger in the bore than the latter. If any are willing to look into the results of the drainage on our Central Park, the most stupendous work of the kind in this country, and one of the best conducted, they will find that the one and a-half-inch and two-inch tiles there used for laterals do not run full even after the most violent and protracted rains, and yet from a single 'system' of twelve acres, the discharge, after a recent rain, was at the rate of 3,000 gallons per hour. This error of using too large tile Mr. Johnston fell into, and now that he has learned better, after a twenty years' experience, he cautions his brother farmers against using larger than two-inch tile for laterals. For mains, each farmer must provide as the quantity of water to be conducted is greater or less. In many cases Mr. Johnston has used two rows of four-inch, in others six-inch, and in one, semi-circles of eleven inches, one at top and one at bottom, making a pipe nine inches bore to discharge water. At first, he had many to take up and replace with large pipe to secure a complete discharge. Main drains he makes six to eight inches deeper than those emptying into them—not with an abrupt shoulder, but leveled up, so that the descent may take place gradually in the length of two tiles—29 inches—and always giving the laterals a slight sidewise direction at the end, so that their water will be discharged down stream into the mains.

"Another error he at first fell into was, in having too many drains on lowlands, and not enough on the upland; thus seeking to carry off the effect, while the cause—the out-cropping springs on the hill-side—remained untouched. Where the source of the water is most abundant, the means for removing it should most abundantly be furnished. Rain-water falls on hills, sinks to an impervious stratum, along which it runs until it either finds a porous section through which it can fall to a lower level, or not finding such, continues on the hard bottom to the side of the hill, where it crops out in the form of a spring. If this spring-water is suffered to run down hill, it washes the hill-side more or less, and coming to the lowland, sinks as far as it may into the soil, makes it sodden, and produces bad effects. To drain effectually, then, we must cut off the supply above, and fewer drains will be necessary below. Here is the whole secret of the thing, and here we see why so much money is spent to so little purpose by those who think that they should only drain the wet lowland. Appearances are deceitful, and we should not suppose that a seemingly dry upland is really dry."

Comment on such a character and such a history as this is superfluous. Mr. Johnston's example as a tile-drainer has been of inestimable value to American farmers. As how such a man feeds his cattle and manufactures manure, must be interesting to many, the following additional extract on that subject is given:

"A word as to this most important subject. On poor lands good crops are got by the use of much manure. This all know. But do they know as well that all manure is not equally good; that a cord of it that has been leached by drenching rains throughout fall and winter, and that has been shone upon by the sun through a hundred hot days, has lost the greater part of its efficacy? That the rivulets of brown liquor that run from the barn-yard into the public road will make more wheat than the brown-washed straw which remains? And that, be manure never so well cared for, its value may be increased at will by the food given to the animals that make it? If they don't, Mr. Johnston does; and so, instead of freezing his stock until they are almost *in articulo mortis*, and starving them on dry stocks and refuse hay until the bones well nigh pierce the skin, he has comfortable sheds and deeply-littered yards for his cattle, and feeds them well at regular intervals with sweet hay, oil-cake, bean-meal, and grain. The result—but what other could you expect?—is, that in spring they are in store condition; he loses none, has no disease among them, saves a large quantity of such manure that one cord of it will bring more wheat or corn than four of ordinary dung, and he grows rich. Reader, if you desire to be a good farmer, go and do likewise!"

CHAPTER VII.

Getting the first thousand dollars—How to save—Man wants but little here below—Actual cost of food—Great successes—A dime a day.

ECONOMY is the sheet-anchor of every beginner, no matter what calling he may adopt. Without it, industry and hard work go for almost nothing. As a general rule, men more frequently grow rich from what they save than from what they make. In farming, especially, it may be assumed that this rule has no exceptions. Our actual bodily wants are few, and may be cheaply supplied without converting us into a race of misers. In illustration of these positions I have gathered from various sources some facts sufficiently striking to command general attention, even if they should be found too hard to imitate.

The greatest fortunes have originated in the smallest beginnings. Stephen Girard, the millionaire of Philadelphia, began the world by selling oranges from the head of a barrel in the streets of an obscure country town. His remark in after life was, that when a man had acquired his first thousand dollars, there was no difficulty in becoming rich. John Jacob Astor began his wonderful career of prosperity by buying the skins of skunks and musk-rats. He is reported to have said that it cost him more severe effort to get the first thousand dollars, than all the others.

Mr. Edwin T. Freedley has written much and well on all these subjects. He says, referring to Astor, that,—

"If he had bequeathed to mankind an easy and certain method of overcoming the difficulty, the bequest would have been a far more valuable one than all his fortune; entitling him to the most conspicuous niche in the gallery of the world's benefactors. The task, however, was beyond his powers, as it has proved too vast for abler men. Franklin attempted to teach the true secret of money-catching—the certain way to fill empty pockets—with what success we have seen. Millionaires have favored the world with their dicta and opinions; but the world has not attached any great importance to their sayings, and certainly not been much benefited by their observations. Mankind generally have probably abandoned the idea of discovering a royal road to wealth, and concluded that an individual, or nation, in order to accumulate capital, must earn something by labor, and then save a portion of the product. Something, however, may be done—and a good deal more than has been done—to facilitate this accumulation; to show labor how, without extra exertion, it can increase its rewards; and show economy how, without injury to the physical system, less may be consumed."

Mr. Freedley has gone largely and thoroughly into all the details of the question as to how to get the first thousand dollars. He tells us—

"*First*, How to save. The human mind receives its first practical lessons in the realities of life at a very early period. The child is initiated and instructed in one of the fundamental principles of social science when he discovers that he cannot purchase a cake and also keep his penny—that he must forego the one or part with the other. As a corollary from the proposition, he then comprehends that, to keep his pennies, he must deny himself cakes; and thus, by involuntary deduction, he arrives at a fundamental principle of economy, viz.: *self-denial in expenditures for personal gratification.* The limit to which it is possible to carry this self-denial without injury to health, or diminution of power for production, is somewhat remarkable. The cost of what are absolute and actual necessaries of life is, in most countries, comparatively little—as is evidenced in cases where stern necessity affixes the bounds of possible expenditure. In France, for instance, there are tens of thousands of peasants and of operatives whose daily earnings do not exceed ten cents, and yet they contrive to live gayly on that sum. As a consequence, in no other country has the art of cookery made equal progress. In Paris, an enterprising woman, Madame Robert, furnishes a dinner daily to six thousand workmen for two pence each, her bill of fare being cabbage soup, a slice of boiled beef, a piece of bread, and a glass of wine. In our Southern States, the food of the chief laborers—the men who at one time produced an export value of over two hundred millions of dollars per annum in cotton, sugar, tobacco, and rice—did not probably cost their providers ten cents per day.

"The full allowance for a laboring man and woman—one that toils all the hours of daylight in the field—is a peck and a half of corn meal, and three pounds of fat bacon. In the Cotton States, the average price of the corn is about seventy-five cents a bushel, and the price of bacon eight cents a pound. This would make the week's rations cost fifty-six cents. At still higher rates it would not be a dime a day—in many places, not half that. In many places, though, the negroes do not get half the above rations. We might still further illustrate the principle that the cost of the substances actually necessary for the support of life is small, by reference to the self-imposed abstinence of misers, and the compulsory abstinence of prisoners.

"An item has been circulating in the newspapers, purporting to be the result of some experiments made in a prison, where it was found that ten persons gained four pounds of flesh each in two months, eating, for breakfast, eight ounces of oatmeal made into porridge, with a pint of buttermilk for dinner, three pounds of boiled potatoes, with salt; for supper, five ounces of oatmeal porridge, with one pint of buttermilk, which cost two pence three farthings per day. Ten others gained three and a half pounds of flesh, eating six pounds of boiled potatoes daily, taking nothing with them but salt. Ten others ate the same amount of porridge and buttermilk, with the potatoes, as the first ten, but for dinner had soup; they lost one and a quarter pounds of flesh each; and twenty others who had less, diminished in size likewise. From this it would seem that potatoes are better diet than smaller quanties of animal food, at least for persons in confinement. The meat-eaters, if they had been allowed ordinary exercise, might have exhibited a very different result.

"A few years ago, a Yankee philosopher of the school of Diogenes, endeavored to ascertain, by actual experiment, how cheaply a man could live; and his experience he has recorded in a volume entitled 'Walden; or life in the Woods.' Mr. Thoreau, the gentleman referred to, being possessed of a capital of $25, took possession of a few acres of land esteemed worthless, and proceeded to erect a cabin by his own labor. The result of his building operations he gives, as follows:

"'I have thus a tight-shingled and plastered house, ten feet wide by fifteen feet long, and eight-feet posts, with a garret and closet, a large window on each side, two trap-doors, one door at the end, and a brick fire-place opposite. The exact cost of my house, paying the usual price for such materials as I used, but not counting the work, all of which was by myself, was as follows; and I give the details, because very few are able to tell exactly what their houses cost, and fewer still, if any, the separate cost of the various materials which compose them:

Boards, mostly shanty boards,............	$8 03½
Refuse shingles for roof and sides,........	4 00
Laths,................................	1 25
Two second-hand windows, with glass,....	2 43
One thousand old brick,................	4 00
Two casks of lime,......................	2 40
Hair,.................................	31
Mantle-tree iron,.......................	15
Nails,................................	3 90
Hinges and screws,.....................	14
Latch,................................	10
Chalk,................................	01
Transportation,.........................	1 40
In all,..................	$28 12½

"'These are all the materials, excepting the timber, stones, and sand, which I claimed by squatter's right. I have also a small wood-shed adjoining, made chiefly of the stuff which was left after building the house.'

"Obtaining his fuel in an adjacent wood, at the cost merely of gathering it, he details his house-keeping expenses as follows:

"'The expense of food for eight months, from July 4th to March 1st, the time when these estimates were made—*though I lived there more than two years*—not counting potatoes, a little green corn, and some peas, which I had raised, nor considering the value of what was on hand at last date, was—

Rice	$1 73½
Molasses	1 73
Rye meal	1 04¾
Indian meal	0 99¾
Pork	0 22
Flour	0 88
Sugar	0 80
Lard	0 65
Apples	0 25
Dried apples	0 22
Sweet potatoes	0 10
One pumpkin	0 06
One watermelon	0 02
Salt	0 03
Amount	$8 74

"It will thus be perceived that his food cost him in money about twenty-seven cents per week. For nearly two years after this, he states, that it consisted of rye and Indian meal (without yeast), potatoes, rice, a very little salt pork and molasses; and his drink was water. The cost of his clothing for eight months he estimates at $8.40¾, exclusive of washing and mending; and his other household expenses, oil, &c., at $2—making his whole expenses for eight months less than $25. 'I learned, from my two years' experience,' he says, 'that it would cost incredibly little trouble to obtain one's necessary food even in this latitude; that a man may use as simple diet as the animals, and yet retain health and strength.'

"Another consoling fact is presented in the fruitfulness of the earth, or in the amount of food that can be produced upon an acre. Nearly every one in our country can command the use of an acre of soil, and let us see how much it is within the bounds of physical possibility to make it produce. Simmonds, in his 'Vegetable Kingdom,' remarks, with regard to the comparative productiveness of crops of human food, that one hundred bushels of Indian corn per acre is not an uncommon crop. One peck per week will not only sustain life, but give a man strength to labor, if the stomach is properly toned to the amount of food. This, then, would feed one man four hundred weeks, or almost eight years. 'Four hundred bushels of potatoes can also be raised upon an acre; this would be a bushel a week for the same length of time, and the actual weight of an acre of sweet potatoes is 21,344 pounds, which is not considered an extraordinary crop. This would feed a man (six pounds a day) for 3,557 days, or nine years and two-thirds. To vary the diet, we will occasionally give rice, which has been grown at the rate of ninety-three bushels to the acre over an entire field. This, at forty-five pounds to the bushel, would be 1,185 pounds; or, at twenty-eight pounds to the bushel, when husked, 2,604 pounds; which, at two pounds a day, would feed a man 1,302 days, or more than three and a-half years.'"

"Such considerations as these are full of consolation to the aspiring, and of encouragement to the very poor. None need despair, and moreover, none need be dishonest. It is possible to accumulate capital, aye, to get the first thousand dollars, from an income not exceeding the most moderate earnings or wages. And let it be inscribed on the lintel of every dwelling—on the desks in every counting-house—on the pericardium of every heart—*It is better to live on ten cents a day than to do a wrong for the sake of money.*

"Again, the economy that leads to wealth implies a *judicious use and profitable investment of savings.* A saving of even a small sum will amount, it is true, within the limits of an ordinary life, to a handsome aggregate; but rapid accumulation in this way can only be attained when money reproduces itself through the agency of compound interest. The wonderful ratio of increase effected by this means, can only be understood by those who have experienced it—though a glimmering of the reality may be obtained by a glance at the following familiar table, interest being calculated at six per cent.:

Savings in 1 Year.	In 10 Years.	In 20 Years.	In 30 Years.	In 40 Years.	In 50 Years.	Savings in 1 Day.
$10	$130	$360	$790	$1,540	$2,900	2¾ cts.
20	260	720	1,580	3,080	5,800	5½
30	390	1,080	2,370	4,620	8,700	8¼
40	520	1,440	3,160	5,160	11,600	11
50	650	1,860	3,950	7,700	14,600	13¾

"The most notable instance that now occurs to me of remarkable success attained through attention to the prompt investment of small sums, is afforded in the annals of Abraham Shriver, of Frederick County, Maryland. With no other resources than a salary of $1,400 a year as judge of a court of inferior jurisdiction, and a small farm of fourteen acres, he succeeded in keeping his personal expenses within the receipts from his farm, which he cultivated like a garden; and by promptly investing his salary every quarter-day—sometimes borrowing for the purpose of anticipating or securing an investment promptly at the time—he accumulated an estate of $150,000. Among the records of Savings Banks, which perform a most useful purpose in collecting and rendering available the dribblets of wealth, no doubt there are many other remarkable instances. In Massachusetts, the deposits in Savings Banks amount to over $23,000,000.

"3. Another element of economy, essential to the accumulation of capital, is *protection against great losses by carefully providing against small ones.* The importance of this principle is thus illustrated by M. Say, a political economist:

"'Being in the country, I had an example of one of those small losses which a family is exposed to through negligence. From the want of a latch of small value, the wicket of a barn yard, looking to the field, was left open. Every one who went through, drew the door to; but having no means to fasten it, it re-opened. One day a fine pig got out, and ran into the woods, and immediately all the world was after it. The gardener, the cook, dairy-maid, all ran to recover the swine. The gardener got sight of him first, and jumped over a ditch to stop him, he sprained his ankle, and was confined a fortnight to the house. The cook, on her return, found all the linen she had left to dry by the fire, burned; and the dairy-maid, having ran off before she tied the cows, one of them broke the leg of a colt in the stable. The gardener's lost time was worth twenty crowns, valuing his pains at nothing. The linen burned and the colt spoiled were worth as much more. Here is a loss of forty crowns, and much pain and trouble, vexation and inconvenience, for the want of a latch, which would have cost three pence; and the loss, through careless neglect, falls on a family little able to support it.'

"Proceeding now to inquire how to *labor with profit*, we remark *first*, that capital is a general term for the accumulated stock of former labor. Its father is labor, and its mother economy. Ties of consanguinity, however, it was long ago discovered, are no preventive against unseemly contention. It is an old proverb, 'When two men ride on one horse, one must ride behind,' but it is not always easy to decide the question of precedence between them. In primitive and unsettled states of society, labor is more powerful than capital. In pruning the luxuriance of na-

ture, and subjugating it to man's uses, the capitalist shrinks into insignificance beside the man of the strong arm and the sharp axe. But as soon as population approaches density, capital vaults into the saddle, and labor must ride on the crupper. In society, as at present developed, especially in the old world, a man who has nothing but ordinary unskilled labor to offer in the market, finds that,

> "'To beg, or to borrow, or to get one's own—
> 'Tis the very worst world that ever was known.'

"Wages would seem to be regulated by the cost of the things supposed to be necessary to support life; and he who would save a portion of his earnings, must reduce his expenditures for living to a very low standard. Nevertheless, there are many well authenticated instances of men who, even in the old world, accumulated some capital from the proceeds of day labor, and eventually became wealthy. How much may be accomplished by an indomitable will—a resolute determination to overcome all obstacles—Foster has illustrated in his "Essay on Decision of Character." He refers to a young man who, having expended a large fortune in prodigality, sat down on the brow of an eminence overlooking what were lately his estates, and there resolved that all these estates should be his again.

"'He had formed his plan, too, which he instantly began to execute. He walked hastily forward, determined to seize the very first opportunity, of however humble a kind, to gain any money, though it were never so despicable a trifle, and resolved absolutely not to spend, if he could help it, a farthing of whatever he might obtain. The first thing that drew his attention was a heap of coals, shot out of a cart on a pavement before a house. He offered himself to shovel or wheel them into the place where they were to be laid, and was employed. He received a few pence for

the labor; and then, in pursuance of the saving part of his plan, requested some small gratuity of meat and drink, which was given him. He then looked out for the next thing that might chance to offer, and went with indefatigable industry through a succession of servile employments, in different places, of longer and shorter duration, still scrupulously avoiding, as far as possible, the expense of a penny. He promptly seized every opportunity which could advance his design, without regarding the meanness of occupation or appearance. By this method he had gained, after a considerable time, money enough to purchase, in order to sell again, a few cattle, of which he had taken pains to understand the value. He speedily but cautiously turned his first gains into second advantages; retained, without a single deviation, his extreme parsimony, and thus advanced, by degrees, into larger transactions and incipient wealth. I did not hear, or have forgotten, the continued course of his life; but the final result was, that he more than recovered his lost possessions, and died an old miser, worth £60,000.'

"In the United States, similar instances of moderate fortunes acquired through persevering industry, and acquired, too, without the sin of covetousness, are so numerous, that a volume would hardly contain them. A leading builder, in New York city, now entitled to a place in the book of the 'Rich Men,' was, some years ago, a bricklayer's laborer, at one dollar per day. He states that out of this sum he always contrived to save fifty cents per day, and laid by $180 the first year. The senior members of many a staunch firm commenced their connection with mercantile life by sweeping out the store in which their fortunes were afterwards acquired. But, notwithstanding the many cheering exceptions to the rule, it is nevertheless true, that ordinary unskilled labor can, at best, make but slow progress toward the accumulation of capital.

"*Secondly.* The rewards of labor and the facility for the acquisition of capital are increased by the possession of some *peculiar knowledge or skill.* A man's pecuniary value may be said to augment in exact proportion to the amount of his effective intelligence, superadded to ordinary physical power. The demand for *educated* labor in progressive countries so far exceeds the supply, that it may, to a certain extent, dictate its rewards. Men, animals, and machines, are everywhere working fruitlessly, or unprofitably, for want of suitable persons to direct their movements; enterprises of the first magnitude languish for want of competent managers; and regions, where nature has been most bounteous in her gifts, are yet comparatively a wilderness, because the arts and mechanism of civilization have not been introduced. The soil of Uruguay, for instance, would produce wheat and Indian corn abundantly and luxuriantly; but its adaptation for the growth of these cereals is rendered comparatively worthless by the absence of suitable mills to grind the products. The sugar-cane of the Southern States, and especially in the Tropics, is wasted immensely, for want of the proper machines and the requisite skill to extract all the sugar from the juice. There are dies in the Indies rarer than the cochineal; fibrous plants more valuable than any flax or hemp; substances more oleaginous than linseed; but they are unappreciated, because the educated mechanism has not as yet prepared them for the world's markets. A quick brain and a ready hand constitute a man Fortune's master. Even women, limited as their opportunities are for gaining a livelihood, independently of being a helpmate to man, wonderfully enlarge the scope of their powers when they combine administrative and manipulative skill. As managers of work-rooms, superintendents, etc., women are especially in demand; and, if qualified, can readily earn from $6 to $12 per week.

"*Thirdly.* Again, the accumulation of capital may be accelerated by *associating with personal labor some responsibility.* Capitalists, in general, are timid, and desire to protect themselves against extraordinary expenses. It is in the nature of capital to surround itself with safeguards, and it willingly pays a premium for guarantees. Thus, though an employer may be able to calculate the cost of an undertaking as measured by the labor involved, he would yet prefer to pay something additional to insure its execution for a definite sum. There are many instances, however, where the skilled laborer alone can form an estimate of the cost; and in such cases he may, by shrewd bargaining, obtain liberal compensation for the work. Many of those who have been remarkably successful in accumulating capital, have done so by advancing, as soon as possible, from the position of simple laborer to that of contractor.

"*Fourthly.* But the lever of greatest efficiency in promoting accumulation is *association of several for a common purpose.* Man, however skillful, is, if unaided by others, a very helpless being. There are tribes, we are told, whose cardinal principle it is for each individual to act independently of his fellows, never helping each other; but their condition, as may be supposed, is but little better than that of the wild animals with which they are surrounded. All of man's most wonderful achievements, those which, if considered disconnectedly from their performance, seem practically and physically impossible, are explained by the mystical power embodied in a combination of numbers for a common purpose.

"And *lastly*, the accumulated stock of the products of labor may be vastly increased by the judicious use of *credit.* Credit is the offspring of good laws and good character. It is one of the advantages of legal protection for person and property that the owners of capital are willing to lend it, trusting to the honesty of the borrower that he will return it, or its equivalent, with rent or interest for its use. It is one of the advantages of good character, and known or presumed punctuality in dealing, that a man may, on his own security, obtain the possession and use of a reasonable amount of capital. Credit, being the representative of capital, performs many of its functions, and confers upon the borrower the same benefits, less the charge for its use. The advantages of credit are nowhere more strikingly exemplified than in the rapid material progress of the United States; and in no other country are the profits from its use so large, when combined with industry and mechanical skill.

"An instance is recorded of a farmer, in Peoria County, Illinois, who lived on a rented farm of eighty acres, for which he paid two hundred dollars rent for the land, and twenty-six dollars for the house; he did all his work himself, except some help in planting corn; had one team of horses; and, after paying his rent and supporting his family, cleared one thousand dollars a year. The Rev. John S. Barger, a Methodist clergyman, in Illinois, furnishes the following interesting account of two Mr. Funks, Jesse and Isaac—no relation of Peter, whose address, as heretofore, is New York city:

"'I will now give you a concise history of the operations of Mr. Funk. Both before and after his marriage, he had made rails for his neighbors at twenty-five cents per hundred. But when the lands where he lived came into market, twenty-five years ago, he had saved of his five years' earnings $1,400, and says if he had invested it all in lands, he would now have been rich. With two hundred dollars he bought his first quarter section, and loaned to his neighbors eight hundred dollars to buy their homes; and with the remaining four hundred he purchased a lot of cattle. With this beginning, Mr. Funk now owns seven thousand acres of land, has near twenty-seven hundred in cultivation, and his sales of cattle and hogs, at the Chicago market, amounted to a little over forty-four thousand dollars in a single year.

"'Mr. Isaac Funk, of Funk's Grove, nine miles distant from his brother Jesse, and ten miles northwest from Bloomington, on the Mississippi and Chicago Railroad, began the world in Illinois at the same time, having a little the advantage of Jesse, so far as having a little borrowed capital. He now owns about twenty-seven thousand acres of land; has about four thousand acres in cultivation; and his last sales of cattle at Chicago amounted to $60,000.'

"In California, farming has yielded equally good returns. A gentleman writes:

"The following facts have come under our knowledge. A German farmer squatted on one hundred and sixty acres of ground, some four years ago. Although he began without a halfpenny, he made in the first year, by wheat growing, the handsome sum of nine hundred dollars, besides paying for his land at one dollar per acre, and for his implements, and buying horses, cows, and oxen, building his house, and completing his fence. For the last two years, his field of forty acres has yielded him 1,100 bushels of wheat per annum, selling for net $1,400; his eggs, poultry, vegetables, fruits, &c., brought in four hundred dollars. He estimates his increase in cattle at eight hundred dollars, and the increase in value of the land at three hundred and twenty dollars. Besides this, according to his own account, he had $2,500 cash in the bank; and, in fact, considered he was worth $10,000, and all this the result of four years judicious labor, single-handed, and commencing totally without capital. A field of 500 acres of wheat has produced, within the last four years, a total of 63,220 bushels, of the value of $108,000.

8*

"An Irish farmer began farming in 1853 with the small sum of $300, made in the mines in company with his nephew, a young lad. He first bought two hundred acres of land, paying a deposit on the same; and the rest of the money was invested in a horse, a cow, and the necessary implements. The first year, his fenced-in fields yielded wheat to the value of $800, which enabled him to pay the remainder of his money for his land, besides repaying him for that expended on his stock. He owns six hundred acres of land, and twenty-eight head of cattle, including seven horses; together with lots of pigs, sheep, and poultry. His arable land is now forty-five acres, besides which he has a large orchard and kitchen garden. In a word, he has made himself a very snug, comfortable home, and something like $4,000 to boot.

"In 1852, an Englishman and two Germans came from the mines, with a united capital of thirteen hundred dollars. They bought six hundred and forty acres of land, and farmed it. Last year one of the Germans sold his share in the increased concern for nine thousand dollars. Some years ago, an intimate acquaintance of ours, a German, in company with another as partner, bought a farm, and took to cultivating it and raising cattle. He now owns upwards of fifteen thousand acres of land, and is worth pretty nearly one hundred thousand dollars. This person, too, began without a halfpenny.

"The Germans are proverbially a frugal, money-making people. One of the Teutons, in reply to a question propounded at the Philadelphia Board of Trade, in relation to discounts, is reported to have revealed the secret of his success as follows: 'I open von grocery, mit cot-fish, and molasses, and onê barrel of viskey. Vell, I goes on und by and by I gets a box of sugar and one box of tea; und by and by I gets a big grocery store mit a box china man in der winder and a horse and wagon to go to market. But I know notin about der book-keeping nor dish-kounts, nor der per cents, but den I tells you vat I knows: *I knows ven I buys sugar for a five cent and sells it for a ten cent, den I makes money.*'

"The cultivation of *fruits* and *vegetables*, especially in the vicinity of large cities, is, if skillfully managed, almost uniformly a profitable business. An acre of superior pear trees has produced to their owner $2,650 in one season. A gentleman who is engaged in cultivating strawberries on ten acres, eight miles from Cincinnati, states that the gross receipts of his patch, in a single season, were $2,210. The expense of picking, including the boarding of the hands, was two hundred and twenty-five dollars, and the expenses of marketing twenty-five dollars. The probable cost of cultivation per annum is fifteen dollars per acre. This gentleman cultivates all his strawberries on new but very hilly ground.

"*Nurseries* generally yield excellent returns for the skill and well-directed labor expended upon them, though, to conduct them successfully, considerable capital is also required. A nursery in the western part of the State of New York, is reported to have made a profit of $80,000 in one year, and another of $20,000. A writer describes a half acre of seedling pears that he saw, as worth, at market prices, ten thousand dollars."

The foregoing appropriate extracts from an article of Mr. Freedley afford the reader a favorable insight to the practical character of his labors. His works on kindred subjects are crowded with facts so pertinent to the design of this volume, that every interested reader may study them with advantage. But instances of individual success, equally striking with any which he has recited, may be found in all sections of our country.

To return to the consideration of the actual cost of living, I give the lively and instructive narrative of Mr. Solon Robinson, of New York, showing how a widow supported herself and four children on a dime a day:

"I had," said she, "one day last week, only one dime in the world, and that was to feed me and my four children all day; for I would not ask for credit, and I would not borrow, and I never did beg. I did live through the day, and I did not go hungry. I fed myself and family with one dime."

"How?"

"Oh, that was not all. I bought fuel, too."

"What! with one dime?"

"Yes, with one dime! I bought two cents' worth of coke, because that is cheaper than coal, and because I could kindle it with a piece of paper in my little furnace with two or three little bits of charcoal that some careless boy had dropped in the street just in my path. With three cents I bought a scraggy piece of salt pork, half fat and half lean. There might have been half a pound of it—the man did not weigh it. Now, half my money was gone, and the show for breakfast, dinner, and supper, was certainly a very poor one. With the rest of my dime I bought four cents' worth of white beans. By-the-by, I got these at night, and soaked them in tepid water on a neighbor's stove till morning. I had one cent left. I bought one cent's worth of corn-meal, and the grocery man gave me a red-pepper pod."

"What was that for?"

"Wait a little—you shall know. Of all things, peppers and onions are appreciated by the poor in winter, because they help to keep them warm. With my meal I made three dumplings, and these, with the pork and the pepper-pod, I put into the pot with the beans and plenty of water (for the pork was salt), and boiled the whole two hours; and then we had breakfast, for it was time for the children to go to school. We ate one of the dumplings, and each had a plate of the soup for breakfast, and a very good breakfast it was.

"I kept the pot boiling as long as my coke lasted, and at dinner we ate half the meat, half the soup, and one of the dumplings. We had the same allowance for supper; and the children were better satisfied than I have sometimes seen them when our food has cost five times as much. The next day we had another dime—it was all I could earn for all I could get to do—two pairs of men's drawers each day, at five cents a pair—and on that we lived—lived well. We had a change, too, for instead of the cornmeal and beans, I got four cents' worth of oatmeal and one cent's worth of potatoes—small potatoes because I could get more of them. I washed them clean, so as not to waste anything by paring, and cut them up, and boiled them all to pieces with the meat and meal."

"Which went the farthest?"

"I can't say. We ate it all each day, and didn't feel the want of more, though the children said: 'Ma, don't you wish we had a piece of bread and butter, to finish off with?' It would have been good, to be sure; but, bless me! what would a dime's worth of bread and butter be for my family? But I had another change the next day."

"What, for another dime?"

"Yes, that was all we had, day after day. We had to live on it. It was very hard, to be sure; but it has taught me something."

"What is that?"

"That poor folks could live a great deal cheaper and better than they do, if they only knew how to economize their food. You have told them how, but they are slow to learn, or loth to change from foolish old practices."

"What was your next change?"

"Oh, yes, I was about to tell you that. Well, I went to the butcher's the night before, and bought five cents' worth of little scrap pieces of lean beef, and I declare I think I got as much as a pound, and this I cut up into bits, and soaked over night—an all-important process for soup or a stew—cooking it in the same water. Then I bought two cents' worth of potatoes and one cent's worth of meal—that made the eight cents; two had to go for fuel every day, and the paper I got my purchases in served for kindling. The meal I wet up into stiff dough, and worked out into little round balls, about as big as grapes, and the potatoes I cut up into slices, and all together made a stew, or chowder, seasoned with a small onion and part of a pepper-pod that I got with the potatoes. It was very good, but it did not go quite so far as the soup either day, or else the fresh meat tasted so good that we wanted to eat more. But I can tell you, small as it may seem to you, there is a great deal of good eating in one dime."

"So there is—what a pity everybody don't know it! What a world of good might be done with a dime!

"Reader, have you got a dime—that is, to spare—only one dime? Give it to that poor widow. Give it? No; you owe it. She has given you twice its value, whether you are one that will feast to-day on a dollar, or be stinted with a dime. She has taught you what you never knew before—the value of one dime."

In his essay on "Economy of Food," Mr. Robinson speaks as follows:

"Meats generally are about three-fourths water; and milk, as it comes from the cow, over ninety per cent. How is it as it comes from the milkman? It is true that chemical analysis does not give us the exact comparative value of food, but with that, and the prices of the various articles, it cannot be a hard matter to determine what is the cheapest or most economical kind of food for us to use. Perhaps of all the articles named, taking into account the price and nutritious qualities, oatmeal will give the greatest amount of nutriment for the least money. But where will you find it in use? Not one family in a thousand ever saw the article; not one in a hundred ever heard of it; and many who have heard of it have a vague impression that none but starving Scotch or Irish ever use it; and, in short, that oats, in America, are only fit food for pigs and horses. It is a great mistake. Oatmeal is excellent in porridge, and all kinds of cooking of that sort, and oatmeal cakes are sweet, nutritious, and an antidote for dyspepsia. Just now, we believe, oats are the cheapest of any grain in the market, and it is a settled fact that oats give the greatest amount of power of any grain consumed by man or beast. This cheap food only needs to be fashionable, to be extremely popular among laborers, all of whom, to say nothing of other classes, eat too much fine-flour bread."

Again, he says:

"Look at the Scotch with their oatmeal porridge, as robust a set of men as ever lived. A Highlander will scale mountains all day upon a diet of oatmeal stirred in water fresh from a gurgling spring with his finger, in a leather cup. Another excellent, though little used, breadstuff, particularly for the sedentary, or persons of costive habits, is cracked wheat, or wheaten grits, as the article is called. That and Graham flour should be used in preference, at the same price per pound, to white flour, because more healthy and more nutritious. One hundred pounds of Graham flour is worth full as much in a family as one hundred and thirty-three pounds of superfine white flour. Cornmeal usually costs less than half the price of flour. It is worth twice as much. It is not so economical in summer, because it takes so much fire to cook it. The first great error in preparing cornmeal is in grinding it too much, and next in not cooking it enough. Cornmeal mush should boil two hours; it is better if boiled four, and not fit to eat if boiled less than one hour. Buckwheat flour should never be purchased by a family who are obliged to economize food. It is dear at any price, because it must be floated in dear butter to be eaten, and then it is not healthy. Oatmeal makes as good cakes as buckwheat, and far more nutritious. But it is more nutritious and is particularly healthy for children, in the form of porridge."

Thousands of the poor, in all ages, have studied the great problem of how to get rich. How to get a farm, as the stepping-stone to wealth, is shown in a new light in the following article from the New York *Tribune* for August, 1859:

"Carlyle has said somewhere that the only state of future torment much regarded or feared now-a-days, is Poverty. How to make money—how to acquire rapidly abundant wealth—is the general and anxious inquiry. Somebody has lately published a book purporting to lay bare the whole art and mystery of making money—including the difficult feat of making the first $1,000—for the paltry sum of one dollar. Fired with emulation, we propose to contribute *our* mite toward the development of auriferous science.

"Let us begin by frankly confessing that we know no royal road to desirable wealth, and greatly doubt the existence of any. We have heard of this or that man making a great pile in a day, or night, or some other short period, by speculation, forestalling, gambling, or something of the sort, but have no faith in that sort of acquisition as either desirable or (save in rare instances) practicable. The Old Book says, 'He that maketh haste to be rich shall not be innocent'—and a more important truth has rarely appeared in any book. If those who are hot on the scent of coffee plantations in Central America or sugar estates in Cuba don't believe it now, ninety-nine in every hundred of them will rue their skepticism before they shall be ten years older.

"Nor can we advise any one to rush to Pike's Peak in quest of the eagerly-coveted gold. A good many are now streaming thither, and more perhaps will follow them, some of whom will probably succeed in their quest, while a far larger number will return poorer than they went, besides being sick, sore and weary. Of the few who make anything in the new Dorado, many more will owe their good fortune to success in gambling or peddling than in personally digging gold. Still less can we counsel any young man to seek a classic education, with a view to eminence in some profession. The professions are all overdone; it would be a blessed thing for all if not another lawyer or doctor should be ground out during the next ten years. The market is already glutted, and the stock held for a better demand is deplorably heavy. Nor do we think it well for even one more youth to addict himself to trade. There are this day as many as two persons engaged in selling goods to each twenty families throughout the country. In other words: Productive industry is paying about one-quarter of its products for the trouble of exchanging them, not taking into account the cost of transportation. If we could reduce our aggregate of merchants of all grades by three-fourths, the remainder might thrive, while selling goods at one-half the profit now charged. And yet we believe the world never afforded larger or better opportunities for acquiring wealth than it does just now; and that there is no better place for trying than our own country affords. Let us give a few hints on this head to those who may need them.

"We will suppose the inquirer to be a young man of fifteen to five-and-twenty, whose educational advantages have been meager, and who is not thoroughly qualified for any field of productive labor. How shall he set about getting rich? We say:

"1. Consider whether you would prefer to be a farmer or an artisan; and, if the latter, of what trade. Having decided, keep your eye steadily on the pursuit you prefer, and find employment in it so soon as possible—doing meantime the best thing that offers, though that be chopping wood at two shillings per cord. Never be idle a secular day when there is any work to be had; and if there is absolutely none where you now are, keep in motion towards a less crowded locality till you find some. Having found work, stick to it heartily and faithfully, and, if it pays you but twenty-five cents per day, contrive some way of living upon twenty.

"Whenever you can find employment in the pursuit you mean to live by, accept it, unless withheld by the necessity of earning more at something else in order to pay your debts. And, in deciding where first to follow so as in time to master the calling you have chosen, prefer the place where you can learn most and fastest to that where you can obtain the largest pay.

"3. Be sure that work and thought go together. Keep your eyes wide open and your mind intent and active. Resolve not only to keep trying till you know how to do everything just right, and then do it no otherwise than that, but to know *why* that is the best way—its reason in the nature of things. If you have chosen farming, be sure to find *some* time in each week to read the best treatises on that noble calling, and keep a keen eye on all the periodicals within reach that treat of it. Take the best one yourself, and study it carefully. In short, give the next two, three or four years to the vital work of mastering your chosen pursuit, so that thenceforth, through every day learning, you may confidently measure your strength in it with any competitor.

"4. Having thus mastered your calling, go to work in it for others for the best wages you can obtain, resolved so to earn them that you will be morally certain to command a larger sum next year. Thus persevere in industry, frugality and temperance, carefully economizing your time and means, until you shall have earned enough to strike out boldly for yourself.

"5. By this time you will have made friends, especially among those of kindred position and habits to your own; and now you can make that sympathy available for your mutual good. Have as many as possible join you in a purchase of land to be divided among you according to your several means and needs; whereby your wealth may be doubled in a month. For example: two or three hundred young men of twenty to thirty, knowing and trusting each other, and each of them a good, thrifty, likely farmer or mechanic, having severally earned and saved from $200 up to $2,000, resolve to buy and settle together. So they send out two or three of their number to look and buy lands for them—in any of the new or of the border Slave States, where even improved lands are cheaper than elsewhere on earth. They select and purchase from 2,000 to 10,000 acres of land, according to the price and their means, survey it into large and small farms and village lots, and sell it at auction to the highest bidder, each member being entitled to buy to the extent of his investment in the purchase, and as much more as he can pay for—each being pledged to settle and improve his tract. The hour this is done and the tract all settled, the members' lands alone are worth double their cost—often much more. The farmers have thus secured lands at wilderness prices, and secured at the same time the vicinage of millers, merchants, mechanics, &c., which gives additional value to lands long since improved; while the carpenter, shoemaker, blacksmith, tailor, tinner, &c., &c., have acquired not merely homes, but life-long customers at the lowest possible prices. CONCERTED EMIGRATION is a plan by which the industrious can at least double their moderate means without making a profit out of anybody else; and there are millions of our people, especially of the young, who might speedily double their little properties by means of it.

"6. Having thus made a home, resolve to spend your remaining days there, and to be one of the best farmers or artisans to be found there or elsewhere. Work steadily but not immoderately; think, observe and read so as to make every blow tell. If your land is mainly timbered, contrive a way to make the timber, if possible, a source of profit; if the soil is rather lean, devote all the time not absolutely needed otherwise to making it richer. Sell only for pay down, and buy likewise for cash. Do not allow your wants to grow faster than your means. Make each mistake or failure a source of instruction and improvement. Form no bad habits—have no liquor on your premises, and no tobacco unless to repel vermin. Have no capital locked up in land that you do not use, unless it be woodland rapidly enhancing in value; nor in fat horses, showy turnouts, nor any sort of fancy property—at least, not till you shall be out of debt, with good buildings, well fenced fields, and everything comfortable about you. Thus move on quietly and steadily; and if you have no bad luck, you may be beyond the reach or fear of want in five years, in comfortable circumstances by the end of ten, and as well off as a man need be within twenty.

"Do you say this seems a slow, humdrum, petty way of getting rich? Well, it is not quite so fast as gambling, or slave-trading, or making $100,000 in a month by cornering an adverse party in the stock market; but let two hundred young men try the course we have so rapidly outlined, against an equal number who try any radically different course—gold-mining, trading, speculating, or the professions—and if our party do not, in the average, come out very far ahead, we shall be forced to conclude that the world is a lottery and that Chance is God."

CHAPTER VIII.

The Long Island Barrens—Their Condition, Price, and Crops.

WITHIN one to three hours' ride of the city of New York, by railroad, there lies a vast body of uncultivated land popularly known as "The Barrens." Why an area so extensive should have remained unappropriated and idle, within cheap and easy reach of a population of nearly a million of consumers, has long been a subject for wonder and speculation. But of late years the public attention has been more particularly turned in that direction, principally by publications of the Farmers' Club of the American Institute, the writings of Hon. John A. Dix, Mr. Thomas Schnebly, Dr. E. F. Peck and others, from whose examinations and reports the substance of the following account has been compiled.

It would perhaps be difficult to say exactly how these lands first acquired the damaging name of "Barrens." In the early settlement of Long Island they secured a very different character, and were held in the highest esteem by all who either described or lived upon them. The writers of two centuries ago referred to them as exceedingly fruitful, with a fine climate, and beautiful streams and bays abounding in all kinds of fish and water fowl. They enumerate the grains, the fruits, and the grasses that they produced. None of them mention the land as being poor or barren. The plains, free from timber, and covered with grass, were wonderful as natural curiosities. A traveller in 1759, says that strangers were always taken to them as the only great curiosity of the kind then known in America. Other portions of the island were covered with immense forests.

It is suggested by Mr. Schnebly that the character of barrens attached to these lands in consequence of their being held in very large bodies, whose owners cultivated only a few acres, allowing the forest to occupy the remainder. The island was mapped out by patents owned by various parties, whose possessions extended from the waters on the north and south sides to the middle and woodlands in the centre. The Nicholas patent, at West Islip, contained originally ten miles square. There were numerous other holders of enormous tracts. They made little effort at cultivation, neither did they desire any others to improve them, "and consequently shut out all investigation, and while they lived among gorgeous scenery, a genial climate, and on so productive a soil, were satisfied with cultivating a few acres to supply their wants, leaving the balance of their territory to unproductiveness, which in time, for that reason, became known as the wild or wood lands of Long Island."

Such is Mr. Schnebly's explanation. More modern times have substituted "barren" for "wild." But the fact of land being thus held in vast tracts on Long Island affords another illustration of the evils inseparable from a land monopoly. These owners became an aristocracy. They not only failed to improve their possessions, but, by refusing to sell, prevented others from doing so. When the monopoly was broken up by their descendants dividing and selling, population flowed in and farms were established. Emancipation in Russia is producing like results, and such will follow the enactment of the Homestead Law.

The great bulk of these lands were comparatively inaccessible to the public. There were roads, it is true, but they were few in number. The island was not a thoroughfare, having crowds of travellers from other States passing over its soil. Few, therefore, saw these tracts, and these few, seeing that they were uncultivated, adopted and propagated still further the popular idea that they were barren. The opening of the Long Island railroad served to dissipate this delusion. It opened up a tract of country ninety-five miles in length.

In September, 1860, the Farmers' Club made excursions over the railroad for the purpose of examining the Barrens. They say that "a stranger unacquainted with the country would readily remark the immense quantity of uncultivated land traversed by the railroad, with only here and there a spot exhibiting tillage, and hence the inquiry would naturally arise, why is it that this extensive tract, so near the great city of New York and its sister city, Brooklyn, remains unsubdued and untilled, and what means can be economically used to make this apparent wilderness productive and remunerative to the labor of man?" Along the railroad line is a district containing 150,000 acres, but partially cultivated. In Queens and Suffolk Counties there were nearly 200,000 acres of unimproved land, as shown by the census of 1855. All this lay within two and three miles of the railroad. In leaving the cultivated lands about Jamaica, there is an unoccupied, uncultivated, unenclosed expanse, without tree or shrub to obstruct the view for miles.

Numerous towns held great tracts containing thousands of acres, which were kept for public use as commons. Hempstead originally held 17,000 acres. North Hempstead disposed of her lands at low prices, and was largely benefited by the influx of new settlers. When the railroad was opened in 1844, it traversed an almost unbroken wilderness, in which scarcely a dwelling was to be seen. But twenty years have wrought wonderful changes in the condition of the lands adjoining it.

In many places on and near the railroad, within about an hour's ride of New York, there is land for sale at moderate prices—Mr. Schnebly says at from ten to thirty dollars an acre, according to location. He refers to the crops produced by various farmers on land in this condemned region. Peach trees grow luxuriantly and bear profusely. Corn, oats, barley and rye are produced as largely as in the best regions, while the average wheat product of the island exceeds that of the State. The State Agricultural Society awarded to Mr. Van Sicklin, of Riverhead, the premium for the best cultivated farm, he having produced crops worth $3,300 at an expense of $1,100 for manure and labor. Another party, who bought land at $12.50 per acre, so cultivated it as to produce one hundred bushels of oats to the acre, and the same season grew turnips which yielded an additional profit of $29 per acre. A tract of these neglected lands has produced twenty-five bushels of wheat per acre, a year or two after being taken up.

These lands have been a frequent subject of discussion at the New York Farmers' Club. As the parties who share in these discussions are experienced agriculturists, some of their opinions are quoted. Professor Nash said:

"It has been stated and denied that the land is loam, and not sand or gravel. I have lately spent some days in examination of this soil, and find that statement correct, and that it is beautifully adapted to garden culture, and capable of producing various crops most profitable to the cultivator. This loam has produced and is able to produce $400 to the acre in strawberries. I wish the slanders that have been spoken against the lands of Long Island could be counteracted, and their value better known and made useful to the world. Although not as rich as prairie soil, it is well worthy of the attention of small farmers and men in search of lands for homes. Such homes can be made upon the wild lands of Long Island as well, to say the least, as in the west."

Dr. Peck said that these lands do not need renovating, but merely cultivation. The whole centre of the island is a natural clover field. He added that upon just such land as that which is called barren, fifty-four bushels of wheat had been grown, taking the State Society's premium. One man had fifty acres of clover in the very midst of the scrub-oak barrens, as fine clover as ever grew. He gave the figures of a thirty-acre farm, which gave a profit for the year of $9,300. There were ten acres in cucumbers. Another farmer raised 4,000 bushels of potatoes, which he sold for $7,000.

A member of the Club having asserted that "it was impossible for any poor man to occupy such land, because he could not improve it," another said in reply:

"The Long Island lands were no poorer than those along the Camden and Amboy Railroad, which have been made the garden spot of New Jersey, and made so by the labor of poor men. He deprecated this continual attack upon Long Island, this constantly telling poor men not to go to that poverty-stricken region to starve. It was this oft-repeated assertion that the lands are barren which keeps them so; it is not because they are so, for it has been proved by the most incontestable evidence that these plains, or barrens, as they are called, can be profitably cultivated. He thought it would prove a great blessing to a great many poor men if they should go out upon the island and cultivate it like a garden. It is no use to talk about capitalists undertaking the work of renovation, if they have got to buy the land and spend a hundred dollars an acre to improve it before they begin to realize a profit. Such men of money are much more likely to spend it in Wall street speculations. For the improvement of Long Island we must look to the laborers, the hard-working poor men, such as the gentleman, in his old-fogy argument, would discourage from the attempt to better their condition."

Another member related an interesting anecdote

of one of his acquaintances, who proved, in the most practical manner, that a poor man could settle upon these so-called poor Long Island lands, and make a good support for his family, and gain property at the same time. He thought it a disgrace to the country and the age we live in to say that these lands were incapable of improvement, except by an expenditure of money so far beyond the reach of all ordinary cultivators that none would be found to undertake the work of improvement.

How some of this land within a few miles of New York is used, and what a variety of products it is capable of yielding, is related in the following lively article from the *Tribune* for July, 1858:

"Long Island is to New York city just what is, or should be, the little inclosure picketed in at the back of every farm-house—the garden-spot furnishing a great abundance of fruits and fresh vegetables to the residents of the mansion in front. Unfortunately, the simile holds good in several respects, for this great garden spot is, like a great many kitchen gardens, run to weeds and waste for lack of care and cultivation. Like the garden divided off in plats, parterres and little nooks, it shows one of lovely flowers and another of weeds—a third is filled with choice fruit, and the next is a nest of wild vines, crabs and brambles—a fourth is waving corn, growing in all the luxuriance of the wonderful soil, while right alongside is a spot that only affords a scant pasture to a stray cow. Instead of being one great garden, unsurpassed by all the world, it is a sad evidence of what neglect and careless cultivation can do to such a spot.

"In the course of our ramble we became satisfied that the soil is capable of furnishing this great city with all its food except, perhaps, the great staples of the west, which alone bear transportation. At Ravenswood we looked into a garden where raspberries are grown by the acre—four or five acres, we should say, in a plat—not for fruit, but to sell the plants to others, at $70 or $80 a thousand, the demand being greater than the owner could supply. Within the same inclosure is an acre or two of rhubarb, which, grown as a crop on several farms on Long Island, yields $500 an acre.

"There are also a great number of sample vines started as stocks for cutting, and to show what is the quality of the fruit of those two new and very superior grapes, the Delaware and Rebecca. The cuttings are all started in thumb pots, in forcing-houses in the winter, and as they make roots, successively removed into larger and larger pots, until there is a mass of fibrous roots filling a gallon pot, which gives a rapid growth to the vine when set out, which buyers greatly prefer to the slower growth obtained from cuttings set in the open ground. The owner is profitably working land that is worth at least $1,000 an acre for building lots.

"Next to his grounds we visited those of a grower of the new cherry currant. But, excellent as that is, he is not satisfied without an attempt to get a better one, and so he has 2,000 seedlings a year old, grown from seeds of the very largest of the fine ones at present grown. Of these, 2,000 plants were grown with much labor, requiring the care and attention of the planter for three years before he can obtain and prove the fruit. He may be rewarded with one choice new seedling, and cast the other nineteen hundred and ninety-nine into the fire, or he may not get even that single one. It is a great labor to grow seedlings until one is obtained better than the original, yet some one must persevere in such labor, or we should never have the choice fruits we now enjoy—such as Hovey's, Hooker's, Longworth's, McAvoy's, Peabody's, Burr's, Wilson's, and other new strawberries; Brinklè's raspberry, Houghton's gooseberry, Lawton's blackberry, and an almost endless list of apples, pears, plums, cherries, and other choice fruits and plants, that somebody has had the patience and perseverance to grow from the seed of old sorts.

"Jumping from Ravenswood—that village of beautiful residences on the bank of the East River, opposite the Isle of Penitence—to East Brooklyn, we shall see as we ride out Division avenue, alias Broadway, a considerable number of small gardens and cultivated spots; but most of the land lies waste and useless to the thousands of starving laborers that throng the streets in pursuit of employment. Not that they are unwilling to work, or the owners of the land unwilling that it should be cultivated, but because the absurd practice prevails of letting cows, horses, hogs, goats, geese, ducks and fowls run at large, pirating their living upon the unfenced lands, and frequently breaking into inclosures. Thus no one can plant a little patch of garden vegetables, which in some cases would save the family from begging or being a public charge. And it is almost an annual charge to fence in a lot, since the material will be stolen for winter fuel, unless closely watched. And so a wide breadth of rich soil, extending as far out as the land has been cursed by city lot surveyors, is a worthless waste, with only here and there a rich green spot, to show us it is not by nature a barren.

"One of these green spots we notice on our left hand, some three miles from Peck Slip Ferry, is a pear nursery, where more than ten thousand trees were in bloom last spring. The most of the trees are grafted upon quince stocks, and are growing vigorously in a clayey loam soil, deeply prepared and highly manured. The trees grown for fruiting are some of them trained to branch three to five stems from the ground, and others one stem, with branches only a few inches from the surface, and top of pyramidal shape a few feet high. Between all these trees plants of Hovey's seedling strawberries are set, and one spot of wonderfully small dimensions, for such a yield, was pointed out, from which forty quarts were picked one evening, which sold readily for forty cents a quart. We should say an acre at the same rate would produce a thousand dollars.

"Between the rows of nursery trees one or more rows of cabbages are grown, by which a clean and continual cultivation is insured to the trees, and a summer crop sufficient to pay for all the labor, leaving the money for trees sold as profit—less the first cost. All of the quince stocks, and many of the pears ready grafted in them, are imported from France. One bill of freight covered 20,000 trees. Who buys all the trees annually imported, or grafted, and cultivated here, we cannot say; but the proprietor assured us that he had constantly upon his book orders ahead of his ability to fill, not being willing to send out any but well-rooted trees.

"Now, all this production from a waste spot has come without premeditation; the proprietor, while engaged in other business, built a house here for his family residence, and could not bear to see all around him a desert of waste; and so he began, first for his own use, to plant pear-trees, and finding his neighbors wanted them, he enlarged his production, until from an amateur he has become a nurseryman, and has made an oasis in the middle of the desert of unoccupied, unfenced city lots, where whole farms have been turned out to common grazing-ground for wandering cows.

"By his side, a man has fenced in with wire several of

these vacant lots, and made a market garden, which is more profitable than twenty times its area of wheat and corn in Illinois. We were pointed to one portion of it that had already yielded two crops this season, worth about $700 an acre, and is now set with celery that will produce four or five hundred more. True, it costs labor and manure, and requires skill beyond that requisite to grow potatoes or pumpkins, but it pays a large profit upon all outlays, and leaves a handsome surplus to reward the man of intellect who does or directs the work.

" A little further on we stopped a few minutes to look at the work of two remarkably skillful English gardeners, father and son, enthusiastic propagators and producers of new plants and rare flowers. Among the curious things in the garden are a thousand thrifty plants of the Lawton blackberry, all propagated from one plant stem last spring, by some secret of their own, which enables them to multiply it almost indefinitely. But the most curious of all things about this garden, where we see everything growing so luxuriantly, is the fact that it is done without manure. They were too poor to buy it, and they cannot afford to grow weeds to make compost, and as the surface had been exhausted by long cropping it in the old style of farming, what were they to do? Go on the same old course of putting nothing on, and taking everything off that the thin surface-soil would produce? No, they could not live by that, and as they would not buy and cart on fertility, they dug for it. They found it two or three feet below the surface, where they put the loose stones, enabling the water to drain off, and roots to run down. Now, when a plant needs fertilizing, a loosening of the surface with a fork lets in the air, and the plant grows again with renewed vigor. Devoted industry and spade labor produce the results we see.

"Next we come to East New York, where waste lots lie

all around, with only here and there a house upon a plain miles in extent. Now, getting beyond these we come to fields of most wonderfully rich and rank green Indian corn, great hay-fields, and smaller ones of rye and wheat stubble, with here and there men digging potatoes. There are but few market gardens or fruit farms, though every one has apple and pear trees for his own use. As a general thing, going on towards Jamaica, farming appears to be done now much as it was in the days of the fathers and grandfathers of the present occupants. Here is an exception. Captain Briggs, who for thirty years followed the sea, and has since been engaged in commerce, and even now, at sixty-five years old, goes every day to the city to attend to business, has still found time to plant an orchard covering twenty-five acres, in which he has sixteen hundred pear-trees, now six to eight years old, and very thrifty, and all in bearing condition. A good many are in fruit this year, although it is a season of general failure, and all show a vigor of growth that proves this fact, that although a man may be bred upon the sea, or has been fifty years of his life engaged in commercial pursuits, it does not disqualify him from cultivating the earth with success if he is a man of sense, who never does a thing without knowing why it should be done.

"For instance, he thought when about to plant his apple-trees, that they were strong-rooted trees, with heavy tops, and should be planted on the northeast side of the orchard to break the prevailing winds from the slower-growing and weaker pear-trees. In looking about, however, he found that apple-trees, as they are usually planted out, are of slow growth, sometimes dying, and sometimes being blown over by the winds that sweep up this gently inclined plain from the broad Atlantic. So he inquired, why? He soon found why. He saw trees stuck down into holes that were so small that the few roots not trimmed off, had to be doubled and coiled around the clump root, and trod and jammed into their narrow quarters, and there the tree was expected to grow.

"'How can it?' said the old sailor. 'These people never reason—I'll have no such work on my farm.'

"It was difficult to get men to work differently, but he could work himself, with his own strong hands. So he had holes dug eight feet broad and two feet deep, and threw back a foot of the top soil in the bottom of the hole, well mixed with compost manure, all as finely pulverized as a garden bed. Then he went to the nursery and bought large trees—'too large to do well,' the nurseryman said, 'he had better take smaller ones.' No, he wanted trees, not whip-stalks. And these trees he wanted with roots, and by determined perseverance he got them with roots. 'Great sprawling things,' the man said who dug them, 'that never could be set out, because nobody was a going to dig holes big enough for that.' He was mistaken, for somebody did dig holes big enough, and somebody got down upon somebody's hands and knees, and 'with fingers weary and worn' straightened out every little fibrous root, and bedded it nicely in the soft earth, and not a tree failed to grow at once; and now, who ever saw such handsome, large, bearing trees at eight years old? Every tree in the orchard stands up as straight as the spars of the owner's ship.

"At Baiseley's Pond we saw men and teams at work digging out the deep bed of muck that for ages has been accumulating, and we said to the man who was digging his potatoes—small potatoes and few in a hill—within a few rods of great piles of this muck, which the contractor had had to buy the privilege of placing on dry land, 'Your crop would be better if you had a good lot of that muck, well rotted and mixed with your soil.'

"'Humph—so I have heard before.'

"And that was his answer to our kindly-meant suggestion; and he bent again over his profitless toil, disinclined to talk of what 'he had heard before.' Some of this muck, compressed like brick while moist, and then dried, burns like cannel coal, leaving an ash highly charged with potash. As we walked over the immense pile, extending in a broad belt around the pond, we picked up tuft after tuft as large as a man's hat, so light that we played football with them. They were made up of a knitted mass of fibrous roots, which would burn like dry wood twigs, and afford an equal per cent. of potash. Yet, while they lie here and decay and waste away, the owners of the poor, sandy soil adjoining will send to Vermont or Western New York for leached ashes, for these they have proved are good for the land, while the value of the muck they have only 'heard tell of before.' They have also heard that it 'is pizen to the land,' and therefore will not listen and learn how to use it and make it most valuable.

"Turning away from this pile of wasted wealth, we drove a mile or two across the plain to visit 'a successful market-truck farmer;' not a 'garden-trucker,' but one who grows field crops for market. To-day he was digging potatoes—'a very fair crop of Mercers, about 150 bushels per acre.' The tops were still very green, and the farmer thought the tubers would have increased about a quarter, if left to ripen, but then they would not bring as much. The price to-day is $3.25 a barrel. The vines are pulled and tubers shaken off between two rows, and the remainder forked out with a five-pronged, flat-tined fork. Then two men pass along picking all the marketable tubers into a basket that holds about three pecks—three being counted to a barrel—saving, as they go along, all the best in a smaller basket, to top off with—the best, of course, always on top. The small potatoes left on the ground are afterwards picked for pig feed, yet sometimes they are sold to the bakers to piece out the superfine flour, and make it carry more water, so as to answer the law that requires bread to be sold by weight. The baskets being filled are loaded upon a wagon that carries forty, with feed and food for a man with two horses, who starts in time to reach the market some time in the night, where he sells his load early the next morning, and returns in time to rest and load up again. This potato ground is sown as soon as cleared of the crop, to wheat and seeded to clover."

From this copious summary of the condition and capabilities of the neglected lands of Long Island, new wonder will be excited at the fact of any large body of them continuing to be unoccupied. The term "barrens" should now become obsolete—they never have been such. The consumer is at their very door, taking not only all they now produce, but ready to devour all that their uncultivated acres could be made to yield. It would be wise for the producer to plant himself beside this consumer. Every man now looking for a farm should go and examine them.

CHAPTER IX.

The neglected Lands of Delaware—Repeopling the Slave Region—Condition, Soil, and Products—Crops and Lumber—Farms for Sale, and Prices—Railroads—Maryland Farms.

THE State of Delaware is rapidly coming before the public in a new and regenerated attitude. The immemorial blight of slavery is fast disappearing. A wholesale Northern emigration is coming in to enlighten and control the remaining heresies which, for a generation at least, must linger among those who were born and educated to believe in them. About eighteen months ago, an association of prominent citizens was formed, with no view to individual profit, but having for its sole object the circulation of knowledge touching the cheap and fertile lands of the State, so that Northern settlers might be drawn thither, thus at once crushing out at the ballot-box the pro-slavery element which had ruled and blasted the region. Many such have become purchasers in consequence of this information, and the number is constantly increasing.

As Delaware presents great attractions for those who desire a farm, much pains has been taken to obtain a full insight into the condition and prices of land, and of its facilities for reaching market. A leading object of the association referred to was the improvement of the State by introducing farmers, artisans, manufacturers, and tradesmen, from other localities. A special object was to develop the agricultural resources of the State, particularly the Central and Southern portions, by encouraging the settlement of truckers, who would purchase and divide the large tracts into small truck farms, producing fruit and vegetables for the Philadelphia and New York markets. This course would not only bring the lands into a higher state of cultivation, but would invite a thriving and enterprising population, greatly adding to the productive wealth of the State, purifying its political atmosphere, and enhancing the value of property.

For the securing of these ends, it was necessary to disabuse the North in regard to the true sentiment of the State, and to assure all who thought of settling there, that they would be welcomed by a large and intelligent portion of the population, from whatever section of the country they might come, and whatever their political views, if loyal to the Government, and disposed to contribute to the development of the resources of the State. The documents thus put forth for general information have been freely used in this chapter.

Such an enterprise appeals strongly to Northern self-interest. It cannot be doubted that, in thus developing Delaware and adjoining sections of the peninsula, with other portions of the South, the trading interests of the North will be largely benefited. By furnishing increased facilities to settlers, capitalists of all classes may make profitable invest ments in new village locations, in lots, and in manufacturing and other establishments, which would greatly appreciate in value with the growth of the place. Destined, as Delaware is, to be the great thoroughfare to Eastern Virginia and the further South, when a general tide of emigration shall set in that direction after the war, investments wisely made in that State must be highly remunerative. In the past, many properties have doubled in value within a few years.

After Delaware has been thus unionized and emancipated, the idea was to present the same potent example of the superiority of free-labor enterprise to other portions of the South. Maryland was to come next, then her neighbors. It is averred, as the most probable hypothesis, that at the close of the rebellion, a large standing army must be maintained for years, or the sentiment of union and liberty must be rapidly created in the South by an infusion of Northern emigration. The blood of the two sections must be made to mingle—the Yankee taking his seat beside the Southron, there teaching him a new lesson of life, how to work or whittle—until he is educated to the right sentiment. Such infusion must be an overflowing one. But the signs of the times indicate that such will be its character.

In Delaware, slavery would still be, as it always has been, a fatal objection to settlement there. The Milford *News* stated, in 1858, that "in Newcastle, the most northerly county in Delaware, where there are scarcely any slaves, improved lands are worth over $50 per acre, while in Sussex, where the bulk of the slaves of the State are held, lands are worth only from $7 to $8 per acre." The same paper says, that three years previous, "A band of 300 Swiss emigrants arrived in New York with all their arrangements made to settle in Delaware. They were farmers, with money to buy land; and hearing that land was cheap in Delaware—a State settled by their fathers—they concluded to settle there; but finding, on their arrival, that Delaware was a slave State, they passed us by, settled in Ohio, and helped to augment the wealth of that young giant of the Union." The sagacious editor declared, eight years ago, that an act of emancipation would at once increase the value of real estate in Delaware five millions of dollars. But emancipation is now at her very door—rebellion has destroyed the institution, and the few slaves are rapidly disappearing. Northern immigration will soon secure an entire eradication of this only remaining drawback to the prosperity of the State. Delaware may therefore be considered a free State, running with Maryland a race for freedom.

The Puritan element having never predominated in Delaware, churches and schools are scarcer than in New England. Slavery has cast its usual blight on the religion and education of her people. Yet these indispensable elements of a high civilization are not wholly absent. They exist in larger extent than in most places where land is equally cheap. They are incomparably better developed than in the new States and Territories of the extreme West. There are churches of different denominations in the principal villages, and a large number within convenient distances in the country. Schools exist in equal number; while the villages and towns contain others of a better class, and sometimes academies. The great need is an infusion of Northern religious and educational elements.

A farmer can do but little active work without health. The country is, in general, pre-eminently healthy, the climate being mild and regular. It is recommended to all who are troubled with pulmonary or bronchial diseases incident to bleak and changeable Northern latitudes. Preachers, who had been compelled by these affections to abandon the pulpit, have been enabled to resume it after a few months' residence in Delaware. It has also advantages, in point of healthfulness, over more newly settled regions, in being less subject to chills and fever. Good soft water abounds, at from ten to twenty feet. All fruits and vegetables can be raised for very early market. A week or two at the beginning of a season, is sometimes worth thousands to the truckman. Ploughing can generally be done all through the winter.

There is almost every variety of soil, most of which may be said to be the natural home of the peach and sweet potato. Major Anthony Reybold has netted $20,000 in one season from peaches. Mr. George Parrish, from 9,000 trees, occupying ninety acres, in 1863, netted $10,000. The tree is here free from the blights that affect it in the North, thus it lives and thrives for many years. Sweet potatoes produce enormously, as much as

200 to 300 bushels per acre. Grapes, melons, and berries of all kinds, produce largely. In the wild blackberry trade alone, almost incredible quantities are gathered by women and children, and sent to market.

Before the construction of the Delaware Railroad, thousands of acres in the central and southern portions of the State were shut in from market. Hence there are vast quantities of virgin land, more or less wooded, whose timber is to be felled, and whose soil developed by the hand of industry. Much timber is still standing along and near the railroad. A profitable business is done in getting out ship timber, while the railroad is constantly requiring wood for fuel and ties for the track. Distant roads are buying ties almost without limit. Numerous tracts of such timber land are for sale. As an illustration of their general character, I copy a single advertisement of a farm of nearly 300 acres, which, finding no purchaser at $25 per acre, was withdrawn from sale:

"With the exception of 50 acres cleared, it is all covered with the heaviest timber to be found anywhere in the State, including heart and yellow Pine, Cypress, Beech, Chestnut, Gum, Poplar, &c., with some scattering White Oak. Most of the large pine timber is perfectly straight, and they run up 60 to 110 feet before the first limb is reached. A great many of them will make from 4,000 to 5,000 feet of lumber each. Many of the cypress trees are from four to six feet in diameter three feet above the ground, and will yield from 5,000 to 8,000 shingles each. The soil of the tract is a deep, black, yellow muck, from three to five feet in depth, thoroughly drained by county ditches, and when cleared will yield from 50 to 100 bushels of shelled corn per acre, and other crops in proportion. On the cleared part there is a small dwelling and outbuildings, and some few fruit trees of various kinds. A good opening for a steam mill."

Thirty such announcements might be quoted, the terms of payment being in all cases extremely accommodating to the buyer. With reference to the timbered land, it may be added that there are numerous mill seats, with flour and saw-mills in operation, and much unused water-power. Many of the latter are for sale at reasonable prices. As fuel is abundant and cheap, so steam power may be and is used to advantage.

The lumber question will be one of interest to many readers. It ranks among the most extensive interests in the country. While in the Free States the forests have been melting away before the axe of the freeman, those in the Slave States have remained comparatively undisturbed. Delaware abounds with tracts of invaluable forests of hard and soft woods, which wait only for the hand of Northern enterprise to lay them low. The lumber could be worked up in many profitable ways. The neighboring market of Philadelphia would consume immense quantities. With an influx of population there will be an increased demand for saw-mills. In new village sites already selected, two or three are needed now. There will also be a demand for planing-mills and sash factories. Turning establishments will pay, as black gum, the best material for carriage hubs, is very abundant. They now leave these woods in rough blocks, to be manufactured elsewhere, instead of being more cheaply turned on the spot where they grew. Spokes and axe-helves could be produced at the lowest cost. Now, Connecticut comes to these Delaware woods and takes away quantities of hickory logs, converts them into spokes and axe-handles, and then brings them back among the very people where the timber grew, and where they could be more cheaply turned. It is the same with manufactories of agricultural implements, with tanneries and other indispensable employments. But one of the greatest wants is a development of the vast deposits of muck which are found in many places. These masses of decayed fertilizers are deposited in basins along the creeks, many feet in depth. But so far they have shared the general neglect. It will be the task, as well as the remuneration, of enterprising settlers, to seize and appropriate these abounding deposits of manure.

Until within a very few years, Delaware has been a comparatively sealed book. It had but one or two railroads. There was no thorough view of the country to be had—no ready ingress, no ready egress. Moreover, it was blasted by dominion of the slave power, and few desired to know what it really was, because none were willing to remove into a slave region. Land was consequently unsalable; but this condition of things has undergone a mighty change. Railroads have been built, which open up to public observation a region heretofore of difficult access to travellers, slavery is certain to be speedily wiped out, and immigrants are pouring in. Even yet the price of land is very low, ranging from ten to forty dollars per acre. Farms with house and outbuildings, fences and fruit trees, within three or four miles of a railroad station, of fair quality and in fair condition, may be obtained for from twenty to thirty dollars per acre. The purchaser can take as much or as little land as he wants. Kent and Sussex counties contain much of this description.

After soil and price have been considered, the vicinity to market is to be looked at. Here it is near and accessible daily by railroad. Products, such as in the West would perish for want of buyers, here find ready sale at high prices. Philadelphia, New York and Boston, are all reached in a few hours. New lines of railroads and steamboats leading direct to New York, are in progress, and when opened will enlarge the present outlets for all kinds of produce. At certain points along the Delaware Railroad there are large establishments for canning peaches, in which hundreds of hands are employed. That portion of the crop which goes to market from the tree, takes the daily train to New York in the afternoon, and reaches that city by daylight the next morning.

In some sections there has already been a marked change in the value of real estate, in consequence of the opening of new railroads, and of the public attention having been directed to these lands. About Middletown, in New Castle county, land which some years ago was thrown out into commons as worthless, cannot be obtained now for less than one to two hundred dollars per acre. Farms in Kent county, now obtainable at from $15 to $40, must within a few years command $75 to $100 per acre. Such are some of the inducements to a settlement in Delaware. If the reader has by this time learned how to get a farm, it is very certain that here he can find one. Population is scarce, land is abundant, and consequently cheap. There are hundreds of owners who, insensible of the advantages they possess, are overstocked with land, and desirous of selling.

The opening of railroads here, as elsewhere, attracted enterprising men from abroad. Among the most active of these is Mr. Alfred T. Johnston, of Milford, in Sussex county, on the Junction and Breakwater Railroad, six miles from Delaware Bay, a hundred from Philadelphia by rail, and at the head of navigation on the Mispillion river. Here the fisheries are very productive, and shipbuilding is extensively carried on. In population Milford ranks next to Wilmington, the second town in the State. Mr. Johnston came from Pennsylvania and settled, five years ago, in Milford. Being shrewd and enterprising, he soon discovered the wants of the region, and the numerous openings for settlement that it presented. He made himself thoroughly acquainted with the land and with those owners who desired to sell, then devoted himself to introducing settlers by making the outside world acquainted with the capabilities of Delaware lands. These were such as to attract hundreds from all parts of the Union. He has introduced so large a population as to change majorities at the polls. In the summer of 1864 I went among some of the settlers thus introduced. Some are mechanics, but most of them are farmers. I saw their improvements, their crops, and questioned them as to their prospects. As most were from a colder climate, they spoke strongly in favor of this. I found none of these recent comers desirous of selling and removing. On the contrary, they were writing letters to former neighbors to invite them to settle beside them. The cheap lands they had bought a few years ago had in all cases risen in value, some double, some treble.

It did not appear that any of these settlers were mere speculators—they came to cultivate the soil. But while doing so, it rose enormously in value. A farmer from Bucks county, Pennsylvania, purchased in 1858 a tract of 475 acres, at $12.50 per acre, within two miles of Milford. It had good buildings and fences, and much of it was cleared. He made payment in a house in Philadelphia, with a mortgage on the farm for the remainder. He has sold off three farms containing 355 acres, in the three, for $12,600, and reserved, during a period of fifteen years, the fruit from 3,000 peach trees he had planted, besides having a farm of 120 acres, now worth $6,000. The peach crop thus reserved is worth $3,000 per annum. Other instances of rapid increase of values were pointed out, some of them quite as remarkable.

Mr. Johnston showed me a long list of properties which he controlled, and was offering to settlers. Many of these were large enough to cut up into half a dozen farms. Others would divide handsomely into two or three, while some were already of the proper size, and others contained only fifteen acres. The aspirations of every possible class of buyers could here be gratified, from the man with a full purse down to him whose whole capital was only a few hundred dollars. Most of these properties were astonishingly cheap, the price, in many cases, being less than the cost of the improvements. All could be had by paying down a fourth, a third, or half the purchase money, with, in most cases, a long term of years for the remainder. Some had been sold without any money being required, a credit having been given for the whole.

For men not strictly farmers, or for farmers with a talent for other business or trade, there were on some of these very advantageous openings for operating in lumber. The timber standing on much of this land could be cut and marketed at a profit of double the first cost of the land, leaving the latter all clear, with a profit besides. This operation has been repeatedly performed, as much of this fine timber stands within a few miles of schooner navigation, and with saw-mills near at hand. In short, for those looking for a new location, there are few regions deserving more attention than the hitherto neglected woods and farms of Delaware.

The lands on which Mr. Johnston is thus introducing settlers, are located principally in Sussex county. It is here the peach tree flourishes in such profitable luxuriance. The product of the county was very large in 1863, but for 1864, it was estimated at 500,000 baskets. This increase is owing to the young orchards, planted by new settlers, coming into bearing. Mr. Johnston's estimate is, that in five years from this time, Sussex county will send millions of baskets to market. The strawberry culture is just beginning on a large scale. But in the common wild blackberry trade, the amount has been so great as to be a most important item in the cash account of every railroad in the State. Sussex county has poured these things into Philadelphia by tens of thousands of bucketfuls, the railroads having opened a market for what formerly perished in the fields. At every railroad station I saw the platform covered with hundreds of buckets of these berries, sometimes a thousand in one place, waiting for the train, while men, women, and children, were constantly bringing in additions to the huge supply. There were buyers from the city who were taking all that came, paying fifty cents per bucket of about eight quarts. In less than twenty-four hours, the great bulk of this supply would be eaten up by the people of Philadelphia.

It will be a subject of wonder with many as to what becomes of this vast supply of light and extremely perishable fruit. The history of a single establishment will go far to remove it. A house in Philadelphia, Messrs. Aldrich and Yerkes, has been several years engaged in the business of canning and preserving fruit. This firm occupies three large five-story warehouses in Letitia street, in which they manufacture pickles, jellies, marmalade, champaign cider, and put up great quantities of tomatoes, strawberries, and blackberries, in cans. These various preserves are sold all over the Union, penetrating even to the gold mines of Pike's Peak, and consumed in every ship that sails the ocean. The demand increases as their productions become better known. They contract for whole orchards of peaches, and last year used 26,000 baskets. Of common wild blackberries they consumed immense quantities: of pine apples 3,000 dozen cans were preserved; of raspberries and currants they consumed wagon loads. In addition to these items, they put up 60,000 jars of honey, and 36,000 bottles of champaign. Pears and good plums they have never been able to procure in sufficient quantity. This year, 1864, they will want some 40,000 baskets of peaches, and fifty acres of pickles. They employ 400 hands, principally women, and can put up nearly 20,000 cans daily. Their pickle tank holds 25 barrels, which are greened in 24 hours, and replaced by as many more. In 1863, this establishment consumed $30,000 worth of sugar.

Here is a single manufactory which buys orchards, cucumber patches, strawberries, raspberries, blackberries, tomatoes, &c., by the acre. But it is only one among many others doing quite as large a business. Every city contains several such, and they are springing up in all the smaller towns. Even in the present infancy of the business, they exercise a marked influence in preventing a discouraging glut of the general market. The public, having had a taste of canned fruits and vegetables, call for increased quantities. The market for these perishable productions, instead of being limited to a few weeks, is made to extend over the whole year. Formerly, a third of the peach crop did not pay for taking to market, and fine fruits of other kinds were frequently given away on reaching the city, the glut being so complete that no buyers were to be found. The canning and preserving establishments are now so numerous as to check these gluts and prevent these losses. From this brief reference to them, the reader will learn something as to what becomes of the enormous amount of the smaller fruits transported over the railroads, as well as of the propriety of going to work at producing them.

Leaving Delaware for Maryland, a very similar condition of things is found to exist. The quantity of land for sale is enormous. The firm of Messrs. R. W. Templeman & Co., of Baltimore, control more than four hundred farms, to which they are inviting the attention of settlers. Some of these contain thousands of acres in a single tract, and could be advantageously divided into smaller farms. Others contain only five to seven acres. Every possible variety of property is embraced in the extensive catalogue which these gentlemen control, while the locations are as various as the different tracts. Many are within easy reach of Baltimore, a city whose daily wants require the products of a large extent of country. In that market, all that the farmer can produce, the fruits and vegetables especially, command highly remunerative prices. Around that city there are farms having fifty acres set with strawberries alone. Some of the first pickings are distributed among northern cities as far as Boston and Portland, but the great bulk of the crop is consumed in Baltimore and Washington. Whoever should be searching for a location in this region will save himself labor, time, and money, by first consulting the gentlemen referred to. They are well acquainted with all these lands, and have been kind enough to furnish me with many facts in relation to them. They have introduced multitudes of settlers from Ohio, Pennsylvania, New Jersey, and the New England States.

The buyer of a farm in Maryland will find no difficulty in arranging favorable terms of payment with the seller. Hundreds of owners are anxious to dispose of their properties. The conditions are as accommodating as in Delaware; for where the competition to sell is so keen, the buyer is very sure to have his own way. I know of farms belonging to a non-resident, who is vainly offering them to Northern men on almost any terms—either to buy, to occupy on shares, or at a nominal rent with privilege of purchasing, or giving them, by some other arrangement, the use of the owner's capital. Many settlers from New Jersey have located in the vicinity of Baltimore, where they are successfully pursuing their former trucking business, feeding both Baltimore and Washington, and doing remarkably well.

By examining the map of Maryland, the reader will perceive that the Chesapeake Bay nearly divides the State into two separate parts, known respectively as the Eastern and Western shore. Beginning at the south end of the Eastern shore, he will find the lands mostly sandy and of sandy loam, up to the line of Talbot county. They are adapted to growing corn, melons, sweet potatoes, and such fruits as peaches, strawberries, &c., and are generally quite level. Where unimproved, they are valued at $10 per acre; where improved, at from $20 to $50.

Continuing north, on the same shore, the lands are generally of a red, stiff soil, level or slightly rolling, and are underlaid with beds of marl. They produce large crops of wheat, corn, fruits, and roots. These lands, unimproved, along the broad waters, are valued at from $30 to $40 per acre; if improved, at from $60 to $80. Further inland, the prices are one-half less. The country along the whole of this shore is indented with numerous broad and deep inlets, furnishing the inhabitants with vast supplies of the finest fish and oysters, crabs, terrapins, and the famous canvas-back and other varieties of ducks. Near the larger towns, and in many other localities, will be found a very superior society. Access to the cities is had by a railroad partly completed, and by steamboats traversing the Chesapeake Bay.

The Western shore will be found, from its most southern point north to the line of Anne Arundel and Prince George county, to be of sandy and sandy loam soil, and in all respects like the first part of the Eastern shore described above, except that the land is more undulating, and produces a light variety of tobacco in large quantities. The value per acre is about the same.

Going North, on the same shore, we reach the

very productive lands of "The Ridge," in Anne Arundel county, and "The Forest," of Prince George county, extending to within forty miles of Baltimore. These lands yield large crops of wheat, corn, tobacco, and roots, are of sandy loam with clay subsoil, are mostly improved, and are valued at $40 to $100 per acre. Continuing North, the lands are of light sandy soil up to the vicinity of Baltimore. When improved, they produce fine crops of early small fruits, melons, peaches, and corn, and are valued, where accessible by railroad, at about $20 to $30 per acre; if not improved, at $12 to $15. The foregoing portions of the State have heretofore been worked exclusively by slave labor, hence the farms will not be found as neat or well cultivated as those in other portions of the State. The inhabitants have access to the cities by steamboat and river craft, the latter furnishing cheap transportation of freight.

From Baltimore, going north and west, to the Alleghanies, and northeast, the lands are high and rolling. The valleys are of limestone, the hills of gray rock, blue slate, and red soil, generally. On the valley lands large crops of wheat and corn are produced, with many cattle. Lime is relied on as the great fertilizer. These lands vary in price with their accessibility. Generally, when distant from the cities, the improved valley lands are valued at from $30 to $80 per acre—the hill lands at about a third less. This region abounds in turnpikes and railroads. Near the large towns, the society is good, and distant therefrom, the people are a thrifty, hard-working class. Here white labor preponderates.

From the base of the Alleghanies, going west, the lands are in many places almost entirely unsettled, and west of the dividing ridge are valued for the large and numerous deposits of the finest bituminous coal, and of hematite and other varieties of iron ore. The table or glade lands of these mountains produce the finest wheat and potatoes, but are too high above tide water for the certain production of corn. Where the mineral region is approached by railroad, the lands are very highly valued; but the table lands can be bought at from $2 to $5 per acre, and offer to a race of hardy settlers very attractive inducements. In this region the winters are long, and sometimes very severe. East of the Alleghanies, down to the latitude of Baltimore, they are shorter and milder. In many parts of the Eastern and Western shore, cattle are not housed except for a very short time.

Northern farmers are astonished at hearing of the low prices at which these Delaware and Maryland farms can be purchased. Compared with rates established throughout the North, they find it difficult to understand why prices should be so high here and so low there. They cannot believe that land in a long populated region, offered so cheaply, can be of any value. If it were, they think it would be quickly occupied by others. The vast quantity for sale is an additional amazement. Yet those who go and see for themselves, discover that they have been mistaken, while many of them purchase and remove to it.

A friend has furnished me with an illustration as to these lands. In the autumn of 1854 he had occasion to visit a gentleman named Seely, living about two miles from Perryville, on the Susquehanna, in Maryland. He there owned and occupied a farm of 200 acres, on which was a stone house standing on a knoll some distance from the main road, and approached by a broad and handsome carriage way. In conversation with Mr. Seely, he stated that he formerly resided in Philadelphia, where he had been engaged in business. He had bought the land five years previously for $15 per acre, or $3,000 for the whole farm.

The buildings, consisting of house, barn, cow-house, wagon-house, &c., were in tolerably good condition at the time of purchase, but the fencing was poor, and the soil almost entirely exhausted by slave labor. The former owner was the possessor of thirty slaves, of various ages and conditions, whose labor had been employed in the cultivation of this farm; but both they and their master had become literally starved out for want of proper management. When the owner sold, he took his slaves and purchase money with him, to seek another locality, there to repeat the same exhausting operation.

At the time of my friend's visit, the whole appearance of the farm had undergone a great and favorable change. Everywhere the fences were in good order, the land had been resuscitated by careful cultivation, and the crops of all kinds were abundant. Mr. Seely stated that his wheat crop alone had that year netted him more than $600. He employed two men the year round, and found their help to be all that was requisite for the proper care of that portion of the farm which was under tillage, except during harvest and haying, when he lent a hand himself. A German servant woman assisted his wife in household duties, and these three hired persons were all the help needed to conduct the farm properly.

My friend said to him, "Mr. Seely, what do you consider this farm worth at the present time?"

He replied, "I have refused $10,000 for it, but I have no disposition to sell. I find farming to be both a pleasant and a profitable employment. I am making money slowly, but surely. Philadelphia and Baltimore afford me a ready market, at good prices, for every thing I can raise, and as I am near a railroad, my transportation is cheap, quick, and easy. After an experience of five years, I love farming, and no ordinary consideration would induce me to leave it for the care, the toil, the anxiety and uncertainties of business in the city."

Mr. Seely continues on his cheaply purchased farm, for which, last year, he was offered $15,000. This case embraces all the strong points which the neglected lands of the two border slave States present. Slave labor exhausts the land, and starves upon it until compelled by a famine of its own creation to emigrate. Free labor comes in and changes the scene from scarcity to plenty. With thirty slaves to work this land, it became too poor to keep them, but with three or four free laborers it yielded bountiful crops, while in ten years its market value increased four hundred per cent.

Northern men, who are thus astonished at the cheapness of these lands, and incredulous as to their value, have overlooked the great underlying fact that the prosperity of these two States has been weighed down by the presence of slavery; that the average value of land in the slave States has uniformly been less than in the free States; that in the former there are no large cities to give value to thousands of surrounding acres, by furnishing markets for their products; that they support no manufactories of their own, but depend almost exclusively on ours; that consumer and producer are everywhere widely separated; that labor, instead of being diversified, is confined principally to agriculture; that instead of being honored, it has been despised; that education and morals have been neglected, and free discussion forbidden. Into communities so governed, Northern men, educated to a higher standard, refused to migrate. It is true, that some were moved to do so, but the census proves that there are more native-born emigrants from the Southern States to the North, than Northerners to the South. Foreigners avoided it for kindred reasons. Thus, with no increase of population from abroad, and with very little at home, it was impossible for land to rise in value. As the West has grown to her colossal proportions by force of immigration, so the South, having none, has failed to increase her numbers to an extent sufficient to enhance the value of her soil.

This unnatural condition of things is now passing away, and a new era is opening on the South. Delaware and Maryland, the two slave States nearest to the North, and, therefore, the most accessible, are already beginning to feel its influence. Slavery removed, they are becoming worthy of Northern attention and enterprise. Delaware is rapidly reviving. Emigration already sets strongly toward her cheap and fertile soil. She is less exhausted than Maryland, and will revive the sooner. An infusion of Northern morals, capital, and enterprise, will regenerate her laws, her institutions, and her habit of thought. Such, also, in the end, will be the happy experience of Maryland. But until the people of the free States enter in by families and colonies, taking possession of the places which nearly two centuries of slavery have made waste, and teaching the inhabitants new thoughts, new habits, and a new civilization, they must remain as they are. Up to this moment they have stood still. If they are to advance, it can only be by help of Northern immigration. As that imperfect form of civilization whose basis was slave labor, has failed to promote State advancement, so the superior one, whose basis has been education and free labor, must be called in to work out in the slave region the only salvation which could prevent it from sinking into a barbarism that was overwhelming the white race as well as the black. Its humanizing influence having been sufficient for itself, it will be found equally potential for others.

CHAPTER X.

Wild Lands of New Jersey—Opening of the first Railroad—Rapid Improvements—New Towns—Hammonton, Egg Harbor City, Vineland, its history, condition, and future—The Neighboring Lands.

Of all the Middle States, none contain so wide an area of uncultivated land, in proportion to the whole, as New Jersey. By the report of the Geological Survey, made in 1856, it appears that of 4,960,595 acres in the State, 3,192,604 acres were, at that time, entirely uncultivated. In 1855, when a bill was before the Legislature for incorporating a company to construct an air-line railroad leading from New York across the lower section of the State, the condition and extent of that uncultivated region were often referred to. The Hon. William Parry, Speaker of the House, made the following statements:

"The amount of land in West Jersey, including the counties of Ocean and Monmouth, which would be benefited by this road, embraces an area of 2,632,000 acres, and in the same section, according to the census of 1850, there are 600,681 acres of improved land, leaving unimproved, mainly for want of railroad facilities, over 2,000,000 of acres. This large extent of country, up to July, 1854, when the Camden and Atlantic Railroad was opened, had no railroad except that skirting along the northern border, following the sinuosities of the river, with spurs to Mount Holly and Freehold, located mainly to accommodate the through travel, without reference to the wants of the interior.

"Can any other State show so large a tract of fertile land, so well adapted to cultivation, and so admirably located as this great peninsula, intercepted between the largest city in the Union and the broad Atlantic, fronting hundreds of miles on the great waters connecting us with Europe, with no more railroads than this section has? The Camden and Atlantic Railroad Company have the credit of opening the way through this heretofore uncultivated portion of our State. Cast your eyes along that road, the location of which is not so favorable for reaching the eastern market as this Air Line, and see the magical effect upon the value of property. Thousands of acres of land, which, previous to its construction, were comparatively of little value, although naturally good, the location being so remote that the price obtained for crops in market would not bear the expense of carting them through the sand, have, since the completion of said road, advanced in value, some, one hundred, some five hundred per cent., and some more, according to the location. The wood which covers most of the high table land, and has heretofore been considered an incumbrance in the way of cultivating the soil, now readily commands from three to four dollars per cord on the road."

The testimony of Mr. Parry becomes especially valuable from two facts—he unites in himself the two professions of land surveyor and nurseryman. He has been for many years the successful proprietor of a nursery embracing two hundred acres, in

Burlington county. As surveyor, he necessarily travelled on foot over the land he describes, and therefore had the fullest opportunity of seeing it, while his lifelong occupation of growing trees and plants of all descriptions, qualifies him as a competent witness as to their capabilities. Thus qualified as an impartial judge, Mr. Parry further said:

"Having spent some time during the past summer surveying in that vicinity, I witnessed what would otherwise have seemed almost incredible; one tract of 30,000 acres was purchased a little before the location of said road, at $1 per acre, and sold shortly after at $5 per acre; $30,000 given and $150,000 received by that transaction, which land is now being divided into small farms, and a large portion of it already sold to actual settlers, at $10 per acre. Another tract of between 20,000 and 30,000 acres has, since the opening of said railroad, been divided into lots and farms, and all sold at $10 per acre to over one thousand purchasers.

"This land has not yet reached one-half its real value, for by this railroad it is brought within one hour's ride of Philadelphia, and it is fertile land, of a sandy loam on the surface, underlaid with clay and gravel, so very essential to retain manures and moisture, and promote the growth of fruit trees, plants and flowers, which flourish remarkably. It is well adapted to raising all kinds of vegetables and grain, which can be taken to market as quick and cheap by railroad as similar articles can in wagons from farms which, owing to their proximity to the city, will bring from $100 to $200 per acre.

"Peaches, which seem to have degenerated in older sections, where the soil has been highly stimulated with artificial manures, there I beheld in a flourishing condition;

trees over fifteen years of age were laden with luscious fruit, bending their slender branches nearly or quite to the ground. There is scarcely an enterprise offering such rich rewards for capital and labor as the extensive cultivation of peaches along the Camden and Atlantic Railroad. Orchards there would rival those so recently celebrated in Delaware.

"Grapes were abundant, and plums without planting—natives of the soil—offered their fruit gratuitously. This is only a part, several other tracts, varying in size from 17,000 to 70,000 acres each, and many of smaller dimensions, are now offered at the low sum of from $5 to $10 per acre, and purchasers and settlers are actually pouring in by thousands, like pigeons to their roost. It seems almost incredible that land of this quality and price should so long remain unnoticed by enterprising men, within thirty or forty miles of Philadelphia, and it is altogether owing to the Camden and Atlantic Railroad that it is now brought before the public.

"Great as these developments are, they dwindle away when compared with what the Air Line will unfold. We who have lived along the Delaware river, and been bound as with a spell to the Camden and Amboy road, cannot appreciate the hidden treasures through the interior of our State. Through this great Peninsula, a large part of which is naturally good land, and all valuable for some purposes, penetrated by lively streams, some of them navigable, many above the limit of navigation would furnish strong power for mills and factories, and all of them, in addition to numerous springs, afford an abundant supply of soft water.

"After the fires, which frequently pass through the woods, destroying large quantities of timber, have abated, Nature, ever active in good works, pushes forth a spontaneous growth of rich, sweet grass, equally nutritious either in the green or dry state, and so protected by dead bushes and trees as to defy the skill of man to gather it; but it is well adapted for feeding stock, and herds of young cattle are marked and driven there from the surrounding country to graze on the pasture, shielded from the sun, and supplied with brooks of pure water, from which they are drawn in the fall or at the approach of winter, fat and ready for the knife.

"I have examined these lands of which I speak, have spent much time in surveying and tracing out the boundaries of hundreds and thousands of acres, and have yet to find the first acre that is not valuable for some purpose—that does not possess intrinsic worth within itself. Some swamps which are termed worthless by the casual observer from a stage *en route* for the fashionable watering places on the coast, when examined through the aid of science, unfold large deposits of iron ore, more than enough for home consumption. Others are well adapted to, and periodically furnish a spontaneous growth of tall, stately white cedars, unequaled for fencing material. Some of the land on which there is no growth of timber or grass, has been called barren, but upon a closer view it is found to consist of a superior quality of glass-sand, and would furnish large quantities for exportation. Other portions are underlaid with large deposits of marl, so very fertilizing to the soil when brought to its surface. The amount of this valuable article is deemed inexhaustible—at least, there is a plenty to enrich the whole State of New Jersey, if we had railroads to distribute it. On other portions, on which there is but a small growth of timber and scanty supply of grass, sand is found to predominate, intermixed with a fine loam. This quality of land is admirably adapted to the cultivation of early vegetables; heavy rains leaving the surface and filtering through the soil, when followed by a hot sun, act like gentle, refreshing showers to hot-house plants; the ground still remains free and mellow to imbibe the atmospheric influence, and does not bake in drying so as to exclude the air, like our heavy, loamy land. This is the reason why sweet potatoes grown on light, sandy soil, are dry and mealy when cooked, light colored and of excellent quality, while those grown on rich, heavy land, worth from $100 to $200 per acre, according to the location, are watery, heavy, dark colored and unpalatable. This is the reason why our light Jersey soil is so very certain for a crop of round potatoes."

Here are some two millions of acres of uncultivated land, shut out from all ready approach, until the year 1854, for want of railroads. In that year the Camden and Atlantic road was opened. It begins at Camden, opposite to Philadelphia, and extends to the ocean at Atlantic City, once a mere barren sand-heap, but now a populous town, with gravelled streets lighted with gas, and built up with great hotels on the beach, and private summer residences of wealthy Philadelphians. At that time Atlantic county contained 315,000 acres, of which only 15,000 were improved; Cumberland contained 335,450 acres, with only 48,460 improved. The railroad traversed an almost desolate wilderness. The land had only a nominal price, and was constantly accumulating in large tracts in the hands of wealthy owners. But no sooner had the railroad been opened than the whole condition of things was changed. Its track is becoming lined with farms and villages. The old growth of pine and scrub oak is being cleared off, buildings erected, lands enclosed, and crops produced. On every side the traveller sees tokens of rapid and substantial improvement, for a market has been opened for whatever the land can be made to yield.

Large tracts have been bought by companies and individuals, and divided into small farms for the accommodation of settlers. It is the West over again, only on a smaller scale. One of the first of these enterprises was at Hammonton, where 5,000 acres were speedily sold in small farms, many of the settlers coming from New England. They find a cash market at Philadelphia for all that they can produce. Yet this was a barren tract, producing nothing salable but wild berries. Another settlement on the railroad is called Egg Harbor City, founded by Germans, who bought a large tract at a low price immediately after the road was opened, and divided it up into town lots and farms. The excellence of the location has attracted to it many families from the West. Nearly all the dwellings are of brick, made on the spot. They have several brick-yards, numerous stores, a printing office, piano factories, saw-mills, and other industrial establishments. The streets are lined with shade trees, and the whole settlement is a model of enterprise, ingenuity and thrift. As seen from the railroad, it will strike every observer as an eminently flourishing place. A great area of wild land has been cleared and farmed, and is producing crops quite satisfactory to the owners. Some are establishing vineyards, others growing tobacco, and others sending great quantities of fruit and vegetables to Philadelphia. All the land thus taken up and improved has quintupled in value within a few years, while that within the neighborhood, though still wild, has advanced in price at least three to four fold.

The natural quality of the soil of this whole section of New Jersey has been long established as good. Its means for improvement are abundant and cheap, in consequence of the vast deposits of marl. You see these deposits cropping out by the road-side, in some places far above the water level. The area of this marl deposit covers 900 square miles, or 576,000 acres, and its benefits are shared by a large district of country lying on each side of it, so that a much greater area than that stated may be fertilized by using it. It has been worth millions of dollars to the State in the increased value of land and produce. It has long been known that Monmouth county potatoes, grown with marl as manure, command half a dollar more per barrel in New York market than any other kind. Mr. Cook's report on the geological and agricultural resources of this portion of the State, made in 1857, informs us, that while the potato crop of Connecticut, New York, and Pennsylvania, has diminished, that of New Jersey and Delaware has largely increased. This increase in New Jersey was mostly in the counties of Monmouth, Burlington, Camden, Gloucester, Salem, and Cumberland, the counties in which marl is found, while in the other counties the gain was small, and in some there was a material falling off. In Delaware, Newcastle county, where marl is found, shows the largest gain. In New York, only three counties showed an increase.

These were on Long Island, and the largest was in those which, there is some reason to believe, lie in the same geological formation with those New Jersey counties in which there was the greatest increase. Uniformly, the potato crop increased most where marl existed. Mr. Cook says:

"The absolute worth of the marl to farmers it is difficult to estimate. The region of country in which it is found has been almost made by it. Before its use the soil was exhausted, and much of the land was so lessened in value that its price was but little, if any, more value than that of Government lands at the West; while now, by the use of marl, these worn-out soils have been brought up to more than their native fertility, and the value of the land increased from fifty to a hundred fold. In these districts, as a general fact, the marl has been obtained at little more cost than that of digging and hauling but a short distance. There are instances, however, in which large districts of worn-out land have been wholly renovated by the use of this substance, though situated from five to fifteen miles from the marl beds, and when, if a fair allowance is made for labor, the cost per bushel could not have been less than from twelve to sixteen cents. Instances are known where it has been thought remunerative at twenty-five cents per bushel."

A ton of marl is sometimes dug from under each square foot of surface; at even half this rate, a square mile will yield nearly 14,000,000 tons. The quantity being thus inexhaustible, the price is consequently low. A right to dig a pit ten feet square costs about $8. The buyer removes the top soil, and plunges into a deposit ten to thirty feet deep, and goes down until stopped by water. In other cases he digs on high ground, and obtains more marl at less cost. The farmer applies from five to twenty-five loads to each acre, according to the exhausted condition of his land. The latter is uniformly benefited by the application. Facts of this description are unanswerable; while the statistics of the potato crop are remarkable enough to secure attention from the most indifferent observer.

Another singular resource for enriching these lands has been discovered and applied. A large crab, known as horse-foot, sea-spider, and king-crab, abounds along the sea-coast, but nowhere in such numbers as on the Delaware Bay. In June, they come on shore in numbers absolutely incredible, for the purpose of depositing their eggs, covering the beach for a distance of forty miles, a moving, crowding, jostling cavalcade, looking as if the beach itself were a living expanse. For weeks they are to be found upon the beach. A million of these unsightly creatures could be gathered in a single mile. The sand is so thickly covered with their eggs, that they are shovelled up by wagon loads, and carried off as food for hogs and chickens. The hogs feed and fatten on the whole crabs, whose average weight is four pounds. During the season when this spontaneous offering of the sea is made to man, the farmers near the ocean come there and collect the living crabs in huge piles, where they cover them with soil, thus composting them into a most energetic manure. A quantity equal to two to four thousand crabs, when applied to the poorest soils, may be relied on to produce 20 to 25 bushels of wheat, though even 30 are by no means uncommon. This fertilizing power has long been known among the farmers.

As it was manifestly impossible to use up the millions of crabs that came upon the beach, within the few weeks that they remained upon it, a manufactory of manure was established at Goshen, in Cape May county, some nine years ago, by Messrs. Ingham and Beesley. These gentlemen first dry the crabs, then grind them into powder, and deodorise the product by combining plaster and charcoal. The manure thus produced they call *cancerine*, and is sold over a wide extent of country. The quantity manufactured in 1863 was about 500 tons. It is cheaper than guano, which contains about one-sixth of ammonia, while this has been proved to contain more than a fourth. On corn, potatoes, &c., this manure produces very marked results, in all cases beneficial. With these bountiful treasures of the sea and land within reach of the people of this portion of New Jersey, it can be the fault of none but themselves if they should fail to convert their admirable soil into one vast garden.

When access to this long neglected portion of the State was opened, in 1854, over the Camden and Atlantic Railroad, nothing was more natural than that the attention of enterprising men from other quarters should be attracted towards it. To secure this attention of the great outside public was a principal object of those who built the road. The holders of the vast tract of cedar, pine, and scrub-oak, through which the road was to run, combined in aid of the enterprise. Half the railroads in our country, all of those which traverse wood and prairie throughout the West, owe their origin to like combinations of great land-owners to open up an inaccessible region to settlement and improvement. Such objects may be denounced as speculative; but their accomplishment has conferred blessings on the Union which cannot be estimated. It has provided new and better homes for millions, built up entire States whose prosperity is an amazement to the world, and done more to secure the perpetuity of the Union than any other movement witnessed among us.

As it has uniformly been in the West on the opening of a new railroad, so it was in New Jersey on the opening of that from Camden to Atlantic City. Enterprising men were drawn to the region thus inviting speculation, investment and improvement. They brought capital, skill and energy, and quickly made an impression. Among the earliest and most thorough-going of these was Mr. Charles K. Landis, of Lancaster, Pennsylvania. This gentleman was impressed with the great value and availability of organized colonization. He secured 5,000 acres on the railroad at Hammonton, and in 1858 his colony was fairly under way. His ideas with respect to colonization appear to have outstripped all others for comprehensiveness, while his plans were definite, practical, and liberal. He sold to none but actual settlers, telling the mere speculators to go elsewhere, and gave especial encouragement to fruit growing. He introduced the New England school system, and kept out the sale of liquor. He laid out streets and roads, and in other ways expended money liberally in promoting the welfare of the settlers. These were of the best class, principally from New England—intelligent, tasteful and industrious. Home manufactures of various kinds were introduced, churches and school-houses were built, good crops were yielded to the farmer, and a general prosperity prevailed which astonished all who witnessed it. The settlement speedily numbered 2,000 persons, who now produce more food than they need, and ship large quantities to New York and Philadelphia.

The experience acquired in settling Hammonton enlarged the views of Mr. Landis, showed him his omissions and mistakes, and gave him ideas which he considered so valuable that he determined to carry them out on a wider field. Accordingly, in 1861 he secured 25,000 acres in one body, in Cumberland county, all in the same wild and uncultivated condition. This tract of waste land lay on the then newly opened railroad from Camden to Cape May, passing through Milville and Glassboro. It covered an area of 45 square miles, with the railroad passing through it, and was within 35 miles of Philadelphia. This settlement he named VINELAND.

In this great undertaking his plan was to establish a perfect, regular, and comprehensive system of public improvement for the benefit of the community to be there located; to found a town in connection with and as an adjunct to an agricultural settlement; to develop therein a system of home manufactures and industry; to promote religion, morals, and a high standard of education, and to provide homes for intelligent and worthy families who might be seeking them.

It was a gigantic project, such as no other individual in this country had ever undertaken to carry out. It required experience, incessant personal attention, great administrative and engineering ability, and the expenditure of a large capital. There have been owners of tracts as large, but none who undertook to transform them from a desolation into a populous community. The lay of this land was such as to admit of its being plotted out as the owner desired. There were no rocks to blast, no mountains to remove, no unwholesome swamps to drain or fill up. He began the enterprise amid the gloom which overspread the public mind immediately after the outbreak of the slaveholders' rebellion. His friends predicted difficulties and discouragements, while all advised him to wait before commencing such an undertaking. But his confidence was not to be shaken—he knew that the very convulsion against which his friends were warning him, was one of those which, of all others, induce men to look for pecuniary safety by purchasing land.

In August, 1861, Mr. Landis went upon his new purchase with a surveyor, for the purpose of locating the first street that was to cross the railroad, since called Landis Avenue. As there was no carriage road either to or through the woods, they traversed the narrow cow-paths afoot, until they reached the spot where the surveyor was to plant his first stake. A profound stillness reigned around them—nothing could be heard beyond a rustling of the leaves—there was not a house within several miles. While the surveyor was planting his stakes, an old dweller among the pines and scrub oaks of that region came up to them, looked at the instruments, and inquired of Mr. Landis what they were doing. He replied that he was locating an avenue a hundred feet wide for a new town, and that within two years he would see the spot they then stood on surrounded with buildings for miles, with farms and orchards where now the forest alone could be seen. The man turned away incredulous, and pitying the infatuation of the projector. No wonder—he had lived seventy years in that particular locality as a wood-chopper, had never been to Philadelphia, did not know how a city looked, and considered the idea of building one in that wilderness as the dream of a lunatic.

But the town was laid out, with many five and ten acre lots, and many farms. Miles of spacious streets and roads were opened, public squares, and a park. Every purchaser was required to plant the front of his property with shade-trees, to build a house within a year, at a certain distance from the roadside, and affording room in front for shrubbery and flowers. Unity of plan was thus secured, insuring the utmost neatness and the highest embellishment—it was to be, in fact, a vast assemblage of beautiful cottage residences. Mr. Landis has already, at his own expense, opened nearly eighty miles of streets and roads, building bridges wherever needed, cleared out acres of stumps and rubbish, established the grade, and on many other improvements expended thousands of dollars in making his great enterprise acceptable to the numerous families who have located on his property.

I visited this remarkable spot in the summer of 1864, to examine its condition and surroundings. I had known and passed over the spot, years before, when it was a perfect solitude, with neither hut nor clearing. It would be impossible, within these limits, to specify the marvellous changes that had been made. The forest had disappeared, and in its place was to be seen a settlement containing some 650 houses and 4,000 inhabitants. There was a rapidly-growing town, having churches, schools, stores, mills, and other conveniences. I conversed with numerous settlers as to whence they came and how they fared in their new location. As a body, they belong to the better class of citizens, are educated, intelligent, moral, and enterprising. The drones which infest other communities are never found in hives like this. Great numbers of them are from New England, while the neighboring States, and even the West, are largely represented in this common centre. Many have built costly and elegant houses. Many are professional fruit-growers and gardeners. Those who buy farms are practical farmers. There are wealthy families in Vineland who remain there because of the mildness of the climate and healthfulness of the place. Taken altogether, the settlement has an old and cultivated look already.

The soil of this great tract varies from a sandy to clay loam, is retentive of manures, and abundantly productive. It produces from 100 to 250 bushels of potatoes per acre, 15 to 25 of wheat, though the premium crop for wheat in Cumberland county, in 1855, was 44 bushels per acre. Of shelled corn, 50 to 75 bushels is the ordinary crop, and two tons of grass. Fruit trees and vines bear abundantly. I saw new peach orchards of thrifty growth, some trees showing fruit, and grape vines giving promise of abundant crops. The winters are so mild as to allow out-of-door work nearly all through them. Mr. Landis told me that for seven years he had not known the ploughing to be interrupted, by reason of frost, for five days in any one winter. All kinds of fruits are cultivated, the five and ten acre lots being mostly devoted to the smaller descriptions. All such are planted so that the picking will come in succession, thus—strawberries, raspberries, blackberries, peaches, grapes, apples, &c.

In driving many miles over Vineland, I entered into conversation with numerous settlers at work by the roadside. Most of these happened to be farmers from the West, from New England, and western New York. All were busy on their growing crops, sometimes in groups of two or three in the corn field. Not one of them but expressed his preference for his new location over the bleak climate he had left. I saw but one desirous of selling and removing, and but one house having on it a handbill as being in market. Most of these farms were just carved out of the woods, showing piles of roots that had been grubbed up. They were, of course, rough looking, like all new clearings in a new country; but the hand of industry was rapidly taming their wildness, and bringing them into prime condition. The general testimony was, that one day's labor on this soil would accomplish twice as much work as if expended on the heavy or strong soil from which they had migrated.

Such was the condition of the farms bought within six months or a year. Those which had been taken up by the first settlers, those of two and a-half years ago, presented a very different appearance. The genial and tractable soil had enabled their owners to work a great transformation even in that brief period. From most of these the stumps had wholly disappeared. Great fields of grain were whitening to the harvest; many acres of peach and apple orchards were to be seen, the former promising to yield a crop the coming season; gardens were full of fine vegetables; the front upon the road had been trimmed up and seeded to grass, while shrubbery and flowers were visible on many of the lawns. Of the thirty-acre farm of Mr. William O. H. Guynneth, a brief notice may serve as an illustration. This gentleman is from Boston, and was among the earliest of the settlers. He bought thirty acres, then utterly wild, now completely tamed. His dwelling-house is so beautifnl a structure as to command admiration anywhere. He has planted orchards, now growing finely, and has acres of excellent wheat. His large cornfield showed as fine a growth as farmer could desire, and so also did his clover crop.

I walked over his ample garden, vineyard, and fruit grounds. Every kind of ordinary garden truck was growing with a luxuriance altogether unexpected, and fully equal to the average of that on lands that sell readily at seven times the cost of his. Several hundred grape vines, Concord, Isabella, and Catawba, two years planted, showed such an excess of fruit as to compel Mr. Guynneth to remove at least half. In no section of New Jersey have I seen the grape vine grow so rampantly as in this ground. Cherry trees, pears, and other fruits, flourished equally well. It was the same with strawberries, gooseberries, and blackberries. This ground had not received a particle of manure. What it now is affords a practical illustration of the real value of this section of New Jersey—three years ago a forest, now the productive and really elegant home of an intelligent and accomplished family.

On reaching the extreme boundary of the Vineland tract, I called on Mr. Robert G. Brandriff, who has here cultivated a farm of 90 acres during the last eleven years. This length of tillage I judged likely to show what was the real stamina of this soil—whether it had any enduring heart in it, or whether it would speedily run down to barrenness. As Mr. Brandriff's land was of even lighter character than that of Vineland, its behavior under long cropping would afford a favorable test for the whole neighborhood. He gave me, without reserve, all the particulars of a truly remarkable history, with permission to use them.

Eleven years ago, this farm was covered with forest. The owner offered it to Mr. Brandriff for $400 for the 90 acres, and an ample time for payment, and being a storekeeper a few miles off, added the important help of a credit on his books for supplies for family use, and materials for buildings, to the amount of $600. At this time Mr. Brandriff was not possessed of a dollar; but he went to work, cleared up his land little by little, a few acres yearly, and thus conquered all difficulties, until now he has 60 acres in cultivation, from which his receipts, in 1863, were $2,000. His family consists of six persons, who have lived well during all this time. His fences and buildings cost him some $1600. He keeps four cows, pigs, and one horse, by which all the work on the easily tilled soil of the farm is done. He hires but one man, except in busy times. For the wants of his family, and the prosecution of other improvements, his annual outlay is $1,000.

Mr. Brandriff showed me his account book for the eleven years he had been at work, in which all his receipts and expenditures were clearly entered, with the balance accurately struck at each year's end. His farm is now worth $6,000, and he has abundant property outside of it to represent any debt he owes. His residence here has not been the hum-drum existence of a mere sandpiper or woodchuck. He is a keen sportsman with line and gun. At the proper season he plunges into the forest that covers much of this section of New Jersey, camps out at night as naturally as an Indian, considers sleep of no consequence when compared with a coon hunt, and is a dead shot at any unlucky deer that crosses his path. The huge antlers hanging up in his shed afford evidence of his skill with the rifle. At other times he visits the neighboring waters of Delaware Bay, where squadrons of wild ducks make generous contributions to his fondness for the gun.

Mr. Brandriff sells his crops at Milville, two miles from his farm. His wheat crop has been 20 bushels per acre, 75 of shelled corn, 200 of round potatoes, 100 of sweet, 560 of carrots, 620 of turnips, while his cabbages pay $100 per acre, and of grass the yield is two to three tons. For manure, his main dependence is on the home product, sometimes using the fertilizers. The particulars of his experience have been thus recited as affording unanswerable evidence of the character of nearly all the land in this heretofore neglected region of New Jersey. Much of it is superior to this particular farm.

The visitor to Vineland cannot fail to notice the absence of fences, even in a ride of fifty miles. No farms have been fenced in, and not a dozen town lots. It had been calculated that $5,000,000 would be required to do the fencing of the whole tract. To save the settlement from this useless tax, Mr. Landis invoked the aid of the Legislature. A new township was erected, bearing his name, in which the running at large of cattle and swine was prohibited—thus each settler fences in his own stock only, and is saved the great cost of fencing out the vicious road-thieves of his neighbors. No other township in New Jersey is found with a similar regulation.

Another peculiarity will be noticed—the total absence of grog-shops, with gangs of loafers congregated about their doors. The law erecting Landis township gave to the people the power of saying whether rum should be sold there or not. So far they have rigidly refused to have it among them, and the character of the settlers coming in will guarantee exclusion in the future. The fine hotel which accommodates strangers has been at no expense for either bar or toddy-stick. These two enactments were portions of Mr. Landis's original plan, and afford satisfactory evidence of the sound morals and practical good sense which he has brought to bear in carrying it out.

No one can spend a day at this place without being strongly impressed in its favor, nor converse with its proprietor without being struck with his remarkable executive capacity. His whole enterprise of settling a tract of forty-five square miles of wild land, has been conceived and carried out on the most comprehensive scale. It is now successfully established on what was three years ago a perfect solitude, by the energy of a single capacious mind. I have seen much of the process of making new settlements on the waste places of the earth; but no instance of methodical planning, of far-seeing judgment, of just calculation of greater ends from a great beginning, than is here exhibited. The original plan, as it was transferred from the projector's mind to paper, can now be seen unfolded in all its symmetrical vastness. Even the details are everywhere visible, all of them in harmony with the whole.

That these results have been actually realized, is shown by the rapid and astonishing success of the settlement. Families are daily coming in from a distance, and selecting homes wherever they think best. As at the beginning, the proprietor continues to convey these locations at low prices and on liberal credit. Mere idle speculators—the men who buy but do not improve, were not wanted, and have been kept out. Many purchasers being well supplied with means, paid cash for what they bought; but to many worthy families the credit given has proved extremely useful. The railroad from Camden through Milville and Glassboro to Cape May, renders the spot accessible to all.

Vineland is probably increasing as rapidly as any new town in the West. In March last, lots were selling so rapidly as to insure the erection of 40 new houses every month, or 480 per annum. No such annual growth as this was realized by William Penn in the early history of Philadelphia. These new buildings are not ephemeral structures, mere shanties to keep off sun and rain, such as one connects with the idea of a new settlement, but substantial and durable houses. Some of them are truly elegant, such only as would be built by men possessing means and taste. When the whole tract has been disposed of, the population of Vineland will be 15,000. Now, the population of the entire county of Cumberland in 1860 was only 22,605; so that in a few years more it will have been nearly doubled by the energy and enterprise of a single individual. Whichever way you turn, progress and improvement of some kind are visible—here a new house is going up, there a new farm is being cleared. The settlement must become in the end an immense fruit garden. Its products reach the two great cities over cheap and rapid railroads, and command cash at generous prices. Its history shows the great public benefit that can be realized from the ownership of a vast tract by one man, when that man uses it and handles it as this tract has been managed. Such wholesale colonization may have been attempted by others, but it has nowhere been so successful as here.

No ducal owner of hereditary acres, either in England or on the Continent, with an annual income greater than the value of the fee of all Vineland, has ever undertaken a similar scheme of colonization. Such men devote their enormous wealth to acquiring more land, not to sharing their acquisitions with their less fortunate neighbors. Instead of clearing up forests and letting in population to improve, and beautify, and acquire permanent and happy homes, they plant the already cleared ground with trees, and shut population out, increasing the difficulty of the masses for acquiring even the smallest freehold. It has been left for a single American citizen, whose capital, unlike that of these baronial landowners, lay more in his head than in his purse, to set before all others thus extensively endowed with land, an example which will add more largely to the sum of human happiness the oftener it may be imitated.

As may be supposed, such a transformation as Mr. Landis has thus effected has powerfully affected the condition and value of thousands of acres within miles around Vineland. Prices have risen, settlers are coming in from abroad, and the area of the great body of waste land is annually becoming lessened by the creation of new farms. The cloud of prejudice which overhung this portion of New Jersey has been effectually dispersed. Railroads have made it as accessible as any other region. Within two hours' ride of it there is a population of a million of consumers whose consumption of its products must annually increase. Within such an atmosphere, these lands, which now sell at from $20 to $30 per acre, must rapidly rise in value, until they reach the prices commanded north of Camden, where, having enjoyed railroad facilities for a longer period, they bring from $100 to $300 per acre. If the reader is at a loss where to find a farm, let him look in this quarter.

CHAPTER XI.

The West—Illinois, and the Central Railroad Lands—Climate, Soil and Productions—Vine Growing in Missouri—Free Lands in the Territories.

THE vast region popularly known as "The West," has been so often travelled by thousands from the older States, and so repeatedly described in print, that all must have a general knowledge of its character and capabilities. Little, therefore, remains for me on these subjects, than a compilation of details appropriate to the matter in hand—where to find a farm.

In the year 1850 Congress granted to the Illinois Central Railroad Company 2,595,000 acres of land, to aid in building a railroad which would open up to sale and settlement a much greater adjoining area belonging to Government, most of which had been many years in market without finding purchasers, even at the low price of a shilling per acre. The quality of the land thus so long for sale was undoubted. It was prairie and rolling land of well ascertained fertility, but, like the long neglected soil of Long Island and of certain portions of New Jersey and Delaware, was effectually shut out from public approach for want of railroads. Mr. Greeley's description of the State, given in 1861, is too graphic to be omitted:

"In the very heart of the great valley, midway between the Arctic and the Tropic, the Atlantic and the Rocky Mountains, lies the State of Illinois, the young Hercules of the West, touching Lake Michigan on the north, and the lower Ohio on the south, with the majestic Mississippi washing her entire western border, and the Wabash skirting her for more than half its length on the east. Her growth, during the last decade, has been really more rapid and considerable than that of any other State, though some of the newest have increased in population by a larger percentage than hers. Her population has all but doubled during the last decade, having risen from some 900,000 to about 1,700,000.

"Other States have each some peculiarity in which it may fairly claim a precedence. Michigan and Wisconsin are both far better timbered, each having an abundance of pine, whereas Illinois has not a stick. Pennsylvania, Virginia, and Missouri, are richer in minerals; Iowa and Kansas have more undulating surfaces, and are (we think) better watered; Ohio lies nearer to the seaboard; New England has her manufactures, and New York her foreign commerce; but in average depth and richness of soil—in capacity to produce, cheaply, grain and grass, meat and vegetables, Illinois is probably the first among the States, and surpassed by no equal area on the face of the globe.

"Originally, scarcity and imperfect distribution of timber, with defective facilities for transportation and travel, were her great drawbacks. Probably three-fourths of her surface were prairie when settlement commenced; while her timber was for the most part stunted and gnarly, by reason of the high winds constantly wrenching, and the fierce fires frequently scorching it. She had no evergreens of consequence, and very few trees from which decent boards could be sawed. Many prairies were ten to twenty miles wide—some were thirty to forty. The deep, black muck which formed the soil was powdered into dust by drouth, or sodden into mire by rain. The moment the prairie sod was cut through, the wheels of each loaded vehicle sank, through half of each year, nearly to the hub; and thus not only building materials, salt and groceries, but fencing and fuel, were to be carted for long distances—a load of wheat being drawn to Chicago, and the proceeds converted into a load of boards for fencing—the journey out and back often consuming a week. Many a load of produce thus marketed, has seen nearly or quite its price absorbed in the inevitable expense of the journey out and in. This impelled the State to engage prematurely in the construction of canals, which involved her heavily in debt, without very materially improving her access to markets. Railroads followed in due season, and did her good service, while, being constructed mainly by private enterprise, whatever advantage accrued to the public was so much clear gain. Still, extensive areas of her soil must have remained unimproved, uninhabited for ages, but for the construction of the Illinois Central Railroad. That great work, munificently endowed with wild lands by Congress, starts from Chicago in the northeast, and Dunleith in the northwest of the State, and converging to a junction near the center, runs thence by a single line to Cairo in the extreme south, at the junction of the Ohio with the Mississippi, the work having thus a total length of over six hundred miles. And, though not run as the profit of the stockholders would have dictated, its course is precisely such as best conduced to the settlement and growth of the State. Millions of acres, else uninhabitable, are by it rendered among the most inviting and valuable of any wild lands on our Continent; and though the Federal grant covered more than two and a half millions of acres, we believe the public domain was increased not merely in value, but in productiveness to the Treasury by this enlightened liberality.

"Illinois, already the fourth, and probably soon to be the third State in the Union—for Virginia is already behind her in every element of consequence and power—is yet in her infancy. Of her soil, probably less than one-fourth has yet been ploughed; and her last crop—immense as it was, especially of corn—is but a fraction of what she can and will produce. We believe her product of this staple already far exceeds that of any other State, while in wheat, beef, and pork, she is scarcely second to any. Her coal is hardly exceeded in abundance by that of any other State; nearly every foot of her surface is underlaid with lime; and her iron, though less abundant, is good. Her chief mart, though hardly thirty years old, ranks seventh among American cities; it promises ere long to be the fifth. Illinois bids fair to have five millions of inhabitants in 1880, and to increase the number to ten millions early in the next century. Her career is hardly begun."

Three years after the Central Railroad Company began their operations, their sales of land amounted to 1,312,373 acres, realizing a total sum of $16,663,823. The terms of sale are probably more liberal than are elsewhere to be found. Had they been otherwise, it would have been impossible to attract to a new and wholly unsettled country the largest body of settlers ever voluntarily collected on one spot within so short a period. The buyer has his choice among a million of acres still unsold, and may take land at from $7 to $12 and upward per acre, according to location. He may pay for it in cash, if able to do so, and thus obtain a discount of twenty per cent.; or he may take land and be allowed four, five, six, and seven years in which to pay for it, but paying the interest yearly in advance. He may buy as small a tract as forty acres, or one as much larger as his means will justify.

The land grant to this Company was the first public gratuity in aid of railroads. When first made, the central portion of Illinois was an unoccupied prairie, as fertile as any soil in the world, but wholly unavailable. It now swarms with population, that along the railroad having trebled within ten years. Great towns have sprung up along its track, and the annual growth of population and wealth is enormous. Here the enterprising man will be sure to find a farm, and the Railroad Company will show him how to get it. Their road is 704 miles in length, and extends from Cairo, in the extreme southern part of the State, to Dunleith, in the northwest, with a branch from Centralia, in the centre, to Chicago, on the shore of Lake Michigan. For all the purposes of agriculture, these lands are equal to any in the world, producing wheat, barley and oats in the north; corn and wheat in the centre; and wheat, tobacco and cotton in the south. In all parts of the State vast numbers of live stock are produced. A healthy climate, a rich soil, and railroads to convey to market the fullness of the earth—all combine to place in the hands of the working man the means of independence. Nowhere can the farmer, the mechanic, the manufacturer, and the laboring man, find surer rewards of industry. With 12,000 common schools, 21 colleges, 48 academies, and a liberal fund for the support of learning, Illinois offers the means of education such as few States can boast. All the conditions favorable to prosperity are to be found here.

From publications made by the Company, most of the facts and descriptions contained in this chapter have been compiled; such, at least, as refer to their lands, and to the statistics of climate and productions. The climate of Illinois is healthy, and the mortality is less than in almost any other part of the country. The immigrant seeking a location regards the healthfulness of the district as a matter of primary consideration, and Illinois, so far as its sanitary condition is concerned, ranks with the most favored States of the Union. The vital statistics collected in 1860 show that in this State the deaths per cent. to the population were in that year only 1.14, while the average of the whole country was 1.27. Extending 380 miles from north to south, Illinois has all the varieties of climate to be found between Boston, in Massachusetts, and Norfolk, in Virginia: in the southern part, the genial climate of Virginia, Kentucky and Tennessee, and in the northern section more nearly resembling that of Pennsylvania, Southern New York, New Jersey and Connecticut.

The soil in the different parts of the State presents very marked characteristics. From the latitude of Chicago as far South as the Terre Haute and Alton

Railroad, the country for the most part is open prairie, with here and there groves of timber, and timbered on the banks of the various streams. The soil in this region consists of a rich black loam, and is remarkably adapted to the production of corn, sorghum and tame grasses. For stock-raising no better land can be found. South of this line the soil is lighter and of a greyish tinge—the country is also more broken, and the timber more plentiful. The small prairies in this region produce the best of winter wheat, tobacco, flax and hemp. From Centralia to Cairo, in the south, the country is heavily timbered. In this district, fruit, tobacco, cotton, and the different productions of the Border States, are largely cultivated and highly remunerative. A large number of saw-mills are erected near the line of the railroad, the lumber from which commands at all times a ready sale.

Indian corn is perhaps the most important crop in the country. It is applied to so great a variety of purposes, and is so indispensable an article for foreign consumption, that however abundantly it may be produced, the constantly increasing demand will press heavily upon the supply. In 1859 the United States yielded 827,694,528 bushels, of which Illinois contributed 115,296,779, about fifty millions of bushels more than any other State. Illinois stands pre-eminently first in the list of corn-producing States.

For the culture of wheat the lands of the Illinois Central Railroad are in all respects equal to any in the State. One great advantage which these lands have, is their nearness to the railroad, by which the purchaser has the means of putting his crop in the market at the earliest or most favorable time, and at a cheap rate of transportation. During the year 1862, the stations on this road sent forward to market 4,688,755 bushels of wheat, besides 567,627 barrels of flour. In Southern Illinois, winter wheat is almost certain to yield a good return to the grower. The reaping, threshing and cleaning machines, now so generally in use, have made wheat-growing a source of great profit to the farmer.

It seems well established that cotton is to become a remunerative crop in the southern part of Illinois. It was cultivated in 1862 in almost every town south of Centralia, and if we regard the planting as an experiment, the result is completely satisfactory. It would be a low estimate to assume that in that year 5,000 bales of ginned cotton were grown. There was a large demand made upon the neighboring States (particularly Tennessee), for cotton seed, and more than one hundred tons had been sent forward from Cairo and distributed.

The rapidly increasing cultivation of sorghum in this country deserves particular notice. In another year Illinois will send to the eastern market thousands of barrels of sorghum molasses, besides retaining sufficient for home consumption. In 1859 this State produced 797,096 gallons, and at that time attention had only just been directed to sorghum. Since then its cultivation has been increased tenfold. A prominent sugar refiner estimates the annual consumption of molasses in the United States at 80,000,000 gallons, and of this vast quantity of sweets it is safe to say the free States consume 60,000,000 gallons. He goes on to say: "This enormous and increasing consumption of molasses and syrups in our Northern States should encourage the western cane growers in their efforts to produce *crops* of western cane syrups, with the certainty that they will find a ready sale for all that will be produced of merchantable quality and in good packages."

Hemp and flax can be produced in Illinois, of as good a quality as any grown in Europe. Water-rotted hemp, from as far north as Sangamon county, when submitted to Government tests, compared favorably with Russian hemp, and exceeded in strength the standard fixed by the Government, in some instances as high as twenty per cent. Good corn lands are good hemp and flax lands, and therefore we may safely conclude that Illinois can produce these important articles much cheaper than they can be imported. If the fabrication of linen goods has made but little progress in this country, it is because the raw material has been grown in but limited quantities. In many parts of the West, farmers have raised flax simply for the seed, and thrown away the fibre as valueless, under the mistaken idea that flax which produced seed could not be worked into fine linen. In the Chicago market, hemp and flax seed are now sold at from three to five dollars per bushel. The Lockport (N. Y.) Flax Cotton Company have contracted with as many farmers of Niagara county as desired to do so, for their crops of flax straw at $10 per ton. In Illinois, with heavy seeding, twenty bushels of seed and three tons of flax straw have been gathered from an acre. This was an extraordinary yield. The *average* crop in Niagara county, New York, in 1862, was one ton of straw and fourteen bushels of seed to the acre.

Much attention is directed to Southern Illinois, on account of its peculiar adaptation to fruit raising. It has the advantage of early season, as well as a soil especially suited to the growing of fruits and vegetables, together with unequaled railroad facilities, by means of which the product is brought to the very door of all the great markets of the Northwest. Fruit placed upon the cars in the evening will reach Chicago the next morning. St. Louis is still nearer than Chicago; and strawberries, tomatoes, &c., are supplied to Cincinnati nearly a fortnight in advance of the ripening of these luxuries in the immediate neighborhood of that city. It is the early market that gives the greatest profit to the fruit grower. Strawberries from Cobden and Makanda are placed in Chicago as early as the 14th of May. The Railroad Company supply every convenience for transporting fruit to market. Cars are run with especial reference to this branch of traffic, and the time of running the trains is so adjusted as best to suit the requirements of shippers. Southern Illinois has become the best fruit-growing region of America. While every part of Illinois is to some extent adapted to fruit culture, it is only in the southern part of the State that all conditions are found in the highest perfection. Pears, apples, peaches, grapes, and strawberries, are produced in all abundance. During the last year, upwards of 200,000 fruit trees were planted in orchards south of Centralia, within six miles of the railroad track: but no matter to what extent they may be multiplied, the demand for fruit will always be in advance of the capacity to furnish what is wanted.

Pork packing has become an immense business in this State, the number of hogs packed in 1862 amounting to 1,484,834 head, half a million in excess of Ohio, which until the last year or two has stood first among the pork-producing States. The following table, giving the number of hogs packed in seven States in 1862, shows a wonderful result:

Illinois	1,484,834
Ohio	981,683
Indiana	587,528
Iowa	403,899
Kentucky	130,920
Wisconsin	196,745
Missouri	284,011
Total	4,069,620

Illinois is the great stock-raising State of the country—sending two thousand head of beef cattle a week to the New York market. In the census return of 1850 the live stock in Illinois had a valuation of $24,209,258, and in 1860 it had increased to $73,434,621—only two States (New York and Pennsylvania) exceeding that amount of value. The raising of stock for market has been the source of many fortunes in Illinois. The Company have large tracts of land well adapted by nature to the raising of cattle, sheep, horses and mules—better adapted, indeed, than are the lands of almost any other State of the Union. During the year 1862, the Illinois Central Railroad brought to Chicago, from various stations along the line, upwards of 30,000 head of beef cattle, and about 10,000 sheep. Wool-growing is a branch of industry that cannot be overdone, and will inevitably be largely increased.

The immense coal deposits of Illinois are worked at different points near the railroad, and thus the settlers are enabled to obtain fuel at the very cheapest rate. Du Quoin and St. John in Southern Illinois, and La Salle, are the principal places from which coal is distributed. The statistics of coal produced in the United States for the year ending June 30, 1860, place Illinois third in the list of coal States—Pennsylvania being first, and Ohio second. In the period named, the coal mined in this State amounted to 14,906,643 bushels, valued at more than a million of dollars. The production at the present time is largely in excess of this amount.

To whatever extent the resources of this State are developed, there can never be any very great accumulation of breadstuffs in this country. It is impossible for Europe to yield enough wheat for its three hundred millions of people, and the soundest writers upon the subject assert that even with the most favorable harvests, three-fourths of the population are inadequately fed. With cheap means of transportation to the shores of the Old World, it is believed that five hundred million bushels of breadstuffs would be annually purchased from the United States. But it is not alone to wheat and corn that the export trade is confined. In Illinois almost every thing that contributes to food for man is produced in excess of the wants of the population, and finds a profitable market in the Eastern States and in Europe.

As before stated, the price of the lands varies from seven to twelve dollars and upward per acre. The prairie lands, except in the immediate vicinity of towns and stations, are generally sold upon the "Long Credit Terms," i. e., payments of principal in four, five, six, and seven years, with interest annually in advance at the rate of six per cent. The timber lands in the southern part of the State, within three miles of the railroad, are sold upon what are known as "Canal Terms," i. e., one-fourth of the purchase money in cash, and the balance in one, two, and three years, with six per cent. interest each year in advance. The following examples illustrate the different terms:

40 ACRES AT $10 PER ACRE—LONG CREDIT TERMS.

	Interest.	*Principal.*
Cash payment....................	$24.00	
Payment in one year.............	24.00	
" two years............	24.00	
" three years	24.00	
" four years............	18.00	$100.00
" five years............	12.00	100.00
" six years.............	6.00	100.00
" seven years..........	....	100.00

40 ACRES AT $10 PER ACRE—CANAL TERMS.

	Interest.	*Principal.*
Cash payment..................	$18.00	$100.00
Payment in one year............	12.00	100.00
" two years............	6.00	100.00
" three years	....	100.00

Thus, forty acres bought on the long credit system, if the credit is all used, will cost the buyer $132 for interest, and $400 for principal—a total of $532. If bought on canal terms, they would cost him $36 for interest, and $400 for principal—a total of $436. If bought for cash, the discount of twenty per cent. would reduce the cost to $320.

Some years ago, Mr. John S. Barger bought five hundred and forty acres of the company's land for $1,513. He had two hundred and eighty acres prepared for seeding, and his gross income, the first year, amounted to $4,428, of which $1,210 was net profit. But to this should be added $1,094, the cost of making the farm, as it was not necessary to repeat the same work another year. Mr. B. had little more than a theoretical knowledge of farming.

Mr. William Waite purchased a prairie farm of eighty acres, in the spring of 1853, paying $4.50 per acre. His land thus cost him $360; the fencing, $400, and the breaking up of sixty acres, $150—a total of $910. None of these outlays would have to be repeated a second year. He marketed 2,100 bushels of wheat and corn, producing him $1,545, leaving him a clear profit of $134, to which the

$910 aforesaid should be added. In three years from the time Mr. Waite first broke up the prairie, his farm was worth $25 per acre.

Every small capitalist who can command only $200 or $300, can quickly acquire a farm in this locality. At first he must put up a shanty of some kind in which to live, then a fence just high enough to turn cattle and horses, these being the only stock permitted to run at large. Then what is known as sod corn may be planted in May, and if the season be fair, it will yield him twenty to forty bushels per acre. The planting is done by striking an axe or spade between the layers of sod, and after dropping the corn, applying the heel of the boot freely. It needs no culture whatever. A man with two horses can ordinarily attend to thirty or forty acres of regular corn land, one ploughing being sufficient. Wheat follows the corn crop. An industrious man can manage eighty acres, by having help at seed-time and harvest.

Mr. W. R. Harris began in 1847 with a capital of $700. He bought one hundred and eighty acres of timber and prairie, of which he broke up fifty-five the first year, and at the end of the fourth year had one hundred and fifteen under the plough. The annual product was $2,000 in cash. At the end of six years, Mr. Harris's capital of $700 had increased to $8,000.

These settlers on the prairie are not subjected to the inconveniences that many will suppose inseparable from a pioneer life. Such as they may be, the robust and thorough-going man will not regard with apprehension. A very neat little house, sixteen by twenty-four feet, a story and a half high, containing five rooms, is furnished and delivered on the cars for about $200. When within one hundred and fifty miles of Chicago, it is put up, plastered, painted, and made ready for occupancy for $350. There are regular manufactories of these portable houses, at which they can be purchased ready made, and the settler fitted out without delay.

But the great West is full of instances like the foregoing. It is true that there have been disastrous failures. Whoever goes there must make up his mind to dispense with some of the comforts and conveniences to which he may have been accustomed. If without capital, he should avoid hanging round the towns, but strike directly for the country, where labor is in demand at paying rates. When able to buy a team, to fence his farm, and pay for a cheap dwelling, then he may safely purchase land. Let him avoid grasping after too many acres at first. The great rock on which many have split is that of seeking to own greater tracts than they can either manage or pay for. Three years of working out will enable a man to save money enough to make a safe beginning. His first crop of sod corn, with a little money, will carry him through the first year, and the second year his land will be mellow enough to bring him a crop of double value.

The Central Railroad Company have given no encouragement to speculators, few of whom are either permanent or improving owners. Their effort has been to secure the actual settler by offering him extraordinary inducements, for it is he whose labors enhance the value of the neighboring lands, and contribute to the traffic of the road. The good effects of this policy have long been apparent. More than a hundred cities and villages now line the railroad, with populations varying from 200 to 10,000 or more, having factories, mills, stores, postoffices, schools, churches, and newspapers. They rapidly increase in numbers and wealth, distributing the comforts and luxuries of civilized life to the settlers, while they open up unlimited opportunities for profitable employment to the business man, the trader, and mechanic.

Other western States afford diversified openings for all classes of enterprising men, whether rich or poor. Kansas has some thirty thousand farms already hewed out of the forest and prairie, on which at least ten millions of dollars have been invested. Some of her towns have grown up as in a single night. Leavenworth expanded, in three years, from a population of 100 to 8,000, with eight newspapers. All the towns of Indiana, Michigan, Wisconsin, and Minnesota, are growing rapidly. The immigrant can find a farm under the Homestead Law, let him look in what direction he may.

Missouri has been extensively settled by colonies of German wine-growers. These form communities by themselves, who are covering the hillsides with vineyards, and have already had remunerating vintages, their wine carrying off the premium at Cincinnati, and coming everywhere into demand. The single town of Hermann, with less than 2,000 inhabitants, produced in one season 80,000 gallons. A vineyard of three to four years old yields the owner two hundred and fifty to three hundred gallons per acre, while a very favorable site has yielded 1,000. One industrious man manages five acres; and as the wine sells for $1.25 to $1.50 per gallon, five acres are sufficient to secure an ample subsistence. The climate is more favorable to vine growing than that of Germany. The German population detests slavery. Now that it has been swept from the soil of Missouri, immigrants are pouring in with every arrival, lands are rising in value, and the Homestead Law is providing thousands of them with permanent homes.

Further west, the territories contain millions of acres of the public domain, all open to settlement by whomsoever chooses to locate upon them. How vast the quantity is, and where situated, will be seen by the following table of acres:

California	94,000,000
Dakota	83,000,000
Nevada	50,000,000
Colorado	66,000,000
New Mexico	72,000,000
Arizona	80,000,000
Utah	62,000,000
Oregon	55,000,000
Idaho	203,000,000
Washington	38,000,000
Nebraska	43,000,000
Kansas	45,000,000

These figures are an approximation to the true amounts, which in all cases are understated.

CHAPTER XII.

Land in the South—Effect of civil war on titles—Progress and results of Pacification—Openings in Louisiana, South Carolina, and Virginia—Great demand for Labor—Cotton Growing—Society after the War.

WHILE the eyes of thousands have for years been directed westward, in search of homes, the slave-holders' rebellion has opened up in that region a new field for enterprising and adventurous spirits from the North and West.

It is a peculiarity of civil war to unsettle or destroy the titles to real estate. A foreign war, even when attended by invasion, produces no such result. When the British overran our northwestern frontier, destroyed Buffalo and other settlements on the Lakes, though personal property was carried off, and houses burned, yet the title to real estate was unimpaired. When they landed on the Chesapeake, sacking Washington and Havre de Grace, holding a possession that was but feebly disputed except at Baltimore, it was personal property alone that changed owners. If the people abandoned their domicils as the enemy approached, they returned to them as he retired. No interlopers having occupied them during their absence, there were none to set up claim to title by possession. The flight of loyal people under such circumstances worked no civil disability. What they suffered was simply a misfortune of invasion. If the enemy had temporarily deprived them of their rights, holding them for the moment in abeyance, yet when he retreated they immediately revived.

But it was not so during the revolutionary contest. That was so emphatically a civil war, that in every State there were two parties in arms against each other. One party fought for American independence, the other for British supremacy, but both were composed of native-born citizens. One was aided by the presence of a British army, the other depended on itself. Had the American people been unanimous in their opposition to Great Britain, it would not have been a civil war, neither could it have been so long maintained against them. But citizen being arrayed against citizen, gave to it a mixed character—it was foreign and civil war combined. To fight for independence was held to be loyal, to oppose it was held to be disloyal.

Those who opposed it were universally known as Tories. Many of them had been office-holders under the king; many of them belonged to the highest classes of society; many were educated, talented, and wealthy; while the fact cannot be disputed, that Toryism was so prevalent that it furnished more armed men to assist in crushing independence, than the Continental army was able to muster for maintaining it. As the Whigs of the Revolution staked their all upon the issue of the contest, the Tories necessarily assumed a like hazard. Many of the latter withdrew from the country at the beginning of the contest, carrying with them their personal effects, but abandoning their real estate. Society was already so disorganized, the future was so uncertain, and money was so universally hoarded, that more were desirous of selling than of buying. Titles had already become uncertain. As the war progressed, the condition of things became worse. Each State enacted confiscation laws designed to cripple the Tories by stripping them of their property. They abandoned lands and houses precisely as the Rebels have been abandoning theirs, and thousands of them never returned to reclaim their possessions. As the American armies advanced, the Tories fled; when the British army moved, others were induced by fear to follow it. The fugitives had no rest. So large a quantity of their real estate was thus brought within reach of the confiscation acts that much of it was overlooked and escaped condemnation and sale.

As the cities and their vicinity had been crowded with Tories, so in those localities their abandoned property abounded. Some of them had been killed in battle, others had fled the country, and dare not return to reclaim what they had left. The few who ventured to do so were again compelled to fly. In multitudes of cases even the voluntary absentees failed to return and resume possession. Titles thus became unsettled. Their properties were entered on by squatters, who eventually gained good titles by long possession. Property of this description is found in all our seaboard cities. Some of it has been thus taken out of its neutral status within a very few years. There are also ground rents which have not been demanded since the first outbreak of the Revolution. Other estates, liable to confiscation, but overlooked at the time, have been squatted on and held until title came of possession. So valuable had some of these become, and so numerous were they in some localities, that sharp lawyers, who devoted themselves to unearthing the secrets of a past era, have grown rich by levying contributions from those who held them in possession, as the price of undisturbed ownership.

These are invariable incidents of civil war. The rich traitor knows beforehand that confiscation of his wealth will be the penalty of his treason. When our population was barely three millions, of whom say only half were hostile to the Government, if civil war resulted so to disloyal owners of real property, what will be the uncertainty and misery among a population nearly six times as large, whose defiant boast has been that they are all traitors? Thousands of them are now passing through the same furnace which consumed the Tories. Like them, having staked all, they have lost all, and are now fugitives in the earth. Others will unquestionably enter into their possessions, the loyal succeeding the disloyal precisely as they did in the last century. The old uncertainties of title may be cured by the action of a Government which seems alive to the necessities of the case. This, if done promptly, will hurry on pacification.

As aforetime, squatting will be practiced everywhere, and in the absence of healing legislation, time will confer title. Owners have disappeared, some killed in battle, some fugitives, others outlawed. Families have been scattered, while others must have perished bodily. Entire States have been made a desolation. Offices of record have been sacked and burned, their parchment contents destroyed or scattered beyond hope of recovery. Trunks full of deeds and wills have gone into the camp fire, or been distributed as military trophies. Fences have been demolished, corner trees cut down, and boundary lines so effectually obliterated, that even the fugitive owners would find it difficult to retrace them, while strangers would find it impossible to do so. The future is full of embarrassment to all titles thus circumstanced. No such wholesale exodus of people occurred anywhere during the Revolution, nor were the armies of that period large enough to produce a tithe of the havoc. The South will thus abound in vacant places which their banished owners dare not return to occupy. The North will rush in to fill them, converting its liberating army into an army of occupation. Into the remaining masses it will infuse new life, new morals, a wholesale education. The sluggards of the South will rise slowly from the depths in which they have been wallowing, conceding to the North, by the logic of events, her true position, that of the ascendant. Subjugated by Northern arms, she will voluntarily acknowledge a new subjugation to Northern mind.

Every Federal army which traverses the Southern

States may be said to be an army of land surveyors. The soldiers see for themselves that the traitorous owners of vast plantations have abandoned them, that repossession is impossible, and that they must fall into new hands. In some cases this change of ownership will be caused by confiscation, in others by a hasty sale at ruinous sacrifice. In other cases the soldiers will acquire title by marriage. All these methods of transition came into active operation soon after the war began. The magnificent climate of the rebellious region, the fertile hills and valleys, soon won the admiration of men whose homes had been among the bleak and rocky soil of New England, or the sparsely populated prairies of the West. Their long stay in that region made them familiar with its value, and there thousands have resolved to settle. With true Northern flexibility of character, they immediately became at home. Intercourse between loyal residents and the soldiers led to intimacy, and intimacy to marriage. Thousands of young volunteers have already married Southern women, and will settle in the South when peace is established. In a single company, as many as thirty such contracts have taken place. These marriages are not restricted to this or that regiment. Wherever the army has gone, there it has been greeted with sympathizers, and there such ties have been established. The longer it remained, and the further it advanced, the more numerous they became. In addition to marriages actually entered into, there will be innumerable engagements to be consummated on the return of peace. Officers as well as men have been equally ready to enter into like relations with Southern women. In the case of wealthy women, some have thus married as much to save their property as to secure a husband.

How far this element of pacification has already progressed, may be seen by the following statement of an army correspondent of the *Tribune.*

"I learn from the most undoubted sources, that in New Orleans, Vicksburg, Memphis, and in smaller places where we have permanent military posts, nothing has become more common than for our soldiers to marry the women of the country. At Memphis, from fifteen to twenty such marriages occur weekly. Let us look at this with the light which social science gives us. Naturally, men establish and live in society. Whatever may be their occupation—fighting, trading, or farming—they will associate in society, and the foundation of this is the family relation. Where opportunity affords, no great length of time can pass in which society will not be organized. It is now from two to three years that we have held most of the places named, and, as a consequence, men without wives, and women without husbands, unite to form a new society, for the old one was, if not destroyed, greatly disrupted, and this, without regard to the fact that formerly they were bitterly opposed to each other. At first widows with large property and no one to attend to it, accepted Union husbands, hoping, and with good grounds, to save their wealth. I know of several young men of good qualities who thus have become rich. I know of one who in a month after marriage, sold $50,000 worth of cotton, which the widow for six months had tried to sell in vain. Other ladies, whose fathers and brothers were in the Rebel army, and had no homes, married; and now, some with fair means and good homes, and who also are well educated, are proud of a gay officer for a husband. There has always been an idea among the women of the South that Northern men make the best husbands. Whether this is true or not is doubtful, but it is certain that multitudes of our officers have every accomplishment, and are skilled in all the arts which attract women. In the ball-room, the parlor, or at promenade, they act the perfect gentlemen; and if the subject of the discourse be literature, art, or philosophy, or even matters of practical life, they are ready to instruct or amuse. So long as we continue to possess the country these marriages will increase. It is evident from this that in a few years more, the Rebel soldiers will be forgotten in the places where they were born, and should they return they will find, beside their houses and land being occupied by others, that their women have become the wives of their enemies."

But whether love or lucre be the motive, the effect on society in that region will be the same. An infusion of Northern sentiment of this wholesale character will be the entering wedge to a sectional salvation. It will be driven home by subsequent accessions of civilians of all professions—mechanics, merchants, traders, lawyers, doctors, speculators. The only true path to the supremacy of civilization is by thus setting free the land, and by planting a true democracy where the most brutalized form of landed aristocracy had been the only ruler. The latter suppressed education, the former will encourage and diffuse it. There will be "a man to every acre, with his right-hand full of brains, and a school-house behind him."

Four years of such progression, even though all were warlike, have already produced a stupendous revolution. Wherever pacification became assured, there agriculture, trade, and commerce, have revived. One Northern manufacturer has received orders from Louisiana for a thousand ploughs, three hundred farm wagons, as many carts, with harness for all, and quantities of tools for carpenters and blacksmiths. Others are crowded with orders from the same region. The makers of agricultural machines, of cotton gins and presses, are equally overrun with orders. Sugar machinery and farm mills cannot be supplied as rapidly as they are wanted.

These are but solitary indications of the astonishing prosperity which is sure to follow in the path of peace. In former times these products of northern workshops were demanded by slaveholders. But the current of events has already changed—that class has disappeared, and a new race of owners and operators has pushed them from their seats. The orders now come from northern capitalists who have succeeded to the abandoned plantations, which they now conduct with great profit by paying to the liberated bondman a fair day's wages for a fair day's work. These openings in Louisiana are numerous, and of all dimensions, small as well as large. An army correspondent thus describes the country above New Orleans:

"For a long distance the road runs in sight of, and but a few miles back from the Mississippi, and passes directly through magnificent sugar plantations—magnificent before the war, but now in many instances tenantless, fenceless, and desolate. The mansions, surrounded with orange groves and with superb shade trees of the live-oak and the cypress, are still there, and the long rows of negro dwellings, far superior to the huts and cabins of Virginia and South Carolina, with their whitewashed sides still glisten in the sun, and look not unlike some neat little village on some Western prairie, clustered around the court-house square. Indeed, all of these plantations are so large, the mansions so princely, and the negro-houses so numerous, that as you whirl by them in the cars it is difficult to realize that you are not passing village after village as you would at the North. But the proud occupants of these princely estates are gone. What they look like as you pass by them in the cars, they are rapidly becoming.

"New England farmers are making them New England villages. The schoolhouse and the church for the first time since these bayous were diked and these vast deltas reclaimed from the Father of Waters and the Gulf, are becoming permanent institutions. In a few years where now you see but a platform alongside of the road for the accommodation of the hogsheads of sugar and the bales of cotton—the crop of the planter through whose estate the road runs—villages will rise as numerous and thriving as along the great Central Road of Illinois.

"Indeed, when this war is closed, and these great plantations are divided up and sold to the soldiers, the sailors, and the German emigrants who will flock to these southern States, Illinois, great as she is, and destined to be much greater, must look out for her laurels. Louisiana could and should have been the wealthiest State in the Union, and when northern industry and enterprise have drained all these swamps and diked all these bayous, no State in the Union can surpass her in the fertility of her soil or in the healthfulness of her climate."

Even in South Carolina, where slavery was most despotic, white superintendence over paid black labor has yielded generous returns to northern capital and enterprise. On thirteen estates at Beaufort, four hundred blacks were employed, rating two children as one hand, and the average wages paid them was fifty-five cents per day. With this help 814 acres were planted with cotton, from which 72,000 pounds of sea-island were obtained, worth $1.50 per pound, while the whole cost was only thirty-eight cents. The poor blacks themselves, when working their own little cotton fields, have lived better than at any former period of their lives, and saved money enough to purchase small farms. When domestic traitors charged that government was wasting millions of dollars in feeding what was said to be a lazy crowd of them at Port Royal, Congress instituted an inquiry, and the Secretary of the Treasury replied that there has been expended for agricultural implements, $77,081; for the purchase of the schooner Flora, $31,350; for white labor, $82,748; for colored labor, $34,527. Total expenses, $225,705. From this expenditure has been realized $726,984. Deducting the above expenses, there remained on hand from this fund $501,279. The Secretary says that no expenditure whatever has been made from the Treasury on account of the cultivation of the plantations or the collection of cotton, or the educational or benevolent care of the laborers. More than half a million of dollars were saved by these operations. If the uneducated negro, just liberated from bondage, can perform these pecuniary wonders, and at the same time acquire a homestead of his own, how brilliant must be the prospect for the educated northern freeman!

It is thus evident that a new chapter is opening in the history of South Carolina. In 1860 rebellion defied the power of the Government. In less than three years the result of this defiance was manifest. The plantation despot became a fugitive. Over 100,000 acres of the sea islands, out of the 19,336,320 acres within the limits of the State, are being rapidly settled by soldiers, sailors, marines, negroes, and teachers who have resided in the Department of the South for six months. The lands have been put up at auction by the Federal Government for non-payment of taxes, and were bought in by its representatives. They are now being re-sold to actual settlers at $1.25 per acre. Each unmarried man or woman can purchase twenty acres, and each head of a family can buy forty acres in addition. So far, some 5,000 acres have been reserved for school purposes, and the dwellings on the portion sold have been appraised at their value, and will be held at that price; the buyers of the soil can purchase the houses only by paying the proper value in addition to the $1.25 per acre for the land. There has been a great rush for the soil thrown open to loyal settlers, and as more of it is offered for sale the demand will increase. A writer on the spot, describing the sale, says:

"It is rather a matter for editorial comment than mere correspondence, that the effect will be to inaugurate the pacification of the country wrested from rebels and added to the domain of the Union. The difficulties which our armies have experienced during the progress of war among hostile populations will, in this department, be prepared for the secessionists, should they ever return to the sea islands. After men have paid for land, and begin its cultivation, they believe it to be truly theirs, and they will fight to defend their rights of ownership. Planting will immediately follow the occupation of lands. Thus a new population of workingmen, white and negro, will take possession of the territory abandoned by the secessionists. Industry and production will follow closely the armies in the field. Every owner of twenty or forty acres, having bought and paid in part at least for his homestead, will feel that he has an interest in his country. The exhaustion of the productive forces that usually remains as the worst evil of a state of war, and that brings with peace a mere syncope of national vitality, will be hereby obviated. The negro population of the sea islands is about 18,000, of whom one-fifth are eligible to pre-empt. In addition to these, the soldiers, sailors, and civilians who are likely to claim land are numerous. It is believed that under the new rule very little land will be left unoccupied. Deeds for lands pre-empted will not be given until the process of cultivation has begun, to prove the good faith of the persons proposing to buy it. Under the guidance of their friends the former cultivators of the soil are staking off their claims. Whole plantations are being settled by families of owners formerly slaves upon the same estates. Generally, the former superintendents of plantations are preparing to settle among their late pupils and subordinate laborers. The place is thrown open to immigration and settlement. At this time, furniture, household utensils, farmers' and mechanics' tools, strong, durable clothing and materials for the wear of both sexes, hardware, and nearly all other useful merchandise suitable to the wants of a new country, are in great demand among all classes at good prices for cash."

As regards that portion of Virginia east of the mountains, the whole region has been made a desolation. Politics and slavery have combined to precipitate rebellion, and rebellion has made it the tramping ground of armies which have eaten out everything but the soil. Every gift of nature seemed to have been lavished on this favorite region—it has an unequalled climate, rivers, coal, and every valuable mineral. But these advantages were thrown away. Virginia employed herself in making Presidents, while New York employed herself in making canals. Virginia was engaged every year in reaffirming the resolutions of '98 and '99, while New York was passing resolutions to build the Erie Canal—not only passing them, but carrying them out. Virginia was employed in setting the machinery of the Federal government in motion, while New York was taxing her colossal energies to start the wheels of commerce. This tells the whole story, and illustrates the superior utility of industrial enterprise over a too long continued devotion to mere political abstractions.

Separated now, the new State of West Virginia, begotten by loyalty from disloyalty, is thus far, in its most prominent features, singularly unlike that of the righteously desolated Old Dominion. The new State is founded upon free labor as opposed to slavery. In that fact alone, she places enterprise, intelligence, skill and forethought, in opposition to their antagonistic elements. Her soil is rich and fertile, as opposed to lands which have been cropped to the last stage of exhaustion by a never-ceasing and never-compensated harvest of tobacco. She has a clear and health-invigorating air, noble mountain-heights, and streams which leap out among them, inviting manufactures. The government is aiming for the greatest good of the greatest number.

Nothing short of a high prosperity was to be expected from conditions such as these. Thus, after disbursing nearly $100,000 since June, 1863, with one half the counties paying no tax whatever, and only a portion of the residue paying their full proportion, there was a balance in the treasury of more than double that amount. This astonishing result is evidence of high prosperity; for if this new State, " wrenched from the grasp of rebellion by military power, and growing on the very hem of civil war, can thus flourish under such circumstances, the restoration of peace and order will develop a prosperity surpassing all that has been written in the history of its ill-starred and traitorous parent."

All through the Southern States there will be innumerable openings for Northern enterprise, whether the adventurers be farmers or mechanics. White labor will be as urgently needed as that of the blacks. The South must continue to grow cotton, rice, sugar, and tobacco, as aforetime. The world needs these products, and is determined to have them. England has relied with pitiable helplessness upon the South for cotton, and France for the maintenance of her stupendous monopoly of tobacco. They have long been chafing under the stoppage of their supplies, and to ensure an early resumption, will at the proper time advance capital without limit to the growers of free cotton. This infusion of new capital will stimulate and invigorate every department of business. More than this, the utter destitution of the South, in consequence of the war, will require the labor of more whites and negroes than she ever possessed, to repair the damages she has suffered.

One half of Kentucky, all Virginia and Tennessee, a large portion of Louisiana, Mississippi, Alabama, and Georgia, have been stripped clean of fencing. Others have suffered vast damage in the same way. When the reader is told that the fences in the single State of New York cost $144,000,000, he may form some estimate of what must be the demand for labor in the South, to make good this single item in the long catalogue of ravages committed by invading armies.

Houses, barns, outbuildings, factories, forges, bridges, have been destroyed by thousands. Other thousands have been stripped of doors, shutters, and siding, to feed the camp fire. In very many instances whole towns have been destroyed. More than half of Charleston lies in ruin. A vast railroad system has been reduced to terrible disorder, and receiving no repairs, will cease to be operated. All these structures were the product of mechanical labor. No one can doubt that the South will in time recover from her desolation, and that these damages will be repaired. But to place her where she stood before the rebellion, will require enormous drafts on Northern industry, not only on that which may remain at home in the workshop, but on that which, to be effective, must migrate to the spot where the ruin exists. In addition to the huge effort of repairing this waste, there will be even more urgent demands for labor. Food must be grown with which to support life, and exportable products from the proceeds of which to purchase clothing, hardware, horses, tools, machinery, and the vast variety of comforts and necessaries which a blockade by sea and land had long excluded.

Those who may be seeking for a farm, will be able in this region to find a hundred. It is doubtful if much capital will be required to secure one. The abandoned properties will be numerous, while of the owners who remain in possession, thousands will be found so impoverished and disheartened as to hail with joy the advent of a Northern coadjutor. It is help that they will need, not land. Of the former they will have too little, of the latter they have always had too much. Such, then, as think that they know how to get a farm, and who have no preference for remaining where they were born, will have little difficulty in finding one somewhere in the South.

On this branch of the subject it will be in point to give the following letter from a prominent officer, dated at Goodrich's Landing, in the northeast district of Louisiana, under date of October, 1863, and addressed to the Hon. Henry T. Blow, of St. Louis:

"I write to you as my friend and as a public man, taking interest in whatever concerns the public good. There is *an immense gold field* down here, and nobody appears to know it. I want it thrown open to the people, so the people can work in it. I refer to the many abandoned plantations from Helena, Arkansas, to Natchez, Louisiana. The owners, most of them, have fled with their negroes to Texas and elsewhere, leaving land that should be occupied.

"During this year, some of the plantations have been worked by Northern men, by hiring negro labor. But few leases were given, as it was late in the season when the idea of cultivation was thought of. Three commissioners were appointed by General Thomas, who gave the leases. The plan was the best that could be adopted on the spur of the moment.

"What leases were given expire in February next, and then I want to see a large laboring population from the North come down here and fill up the country. I lived at Fort Kearney during two gold excitements. One was California, the other Pike's Peak. I saw the great numbers of people that moved there to dig for gold. The gold got there was nothing to what can be made by coming to this country. Let the prospect be advertised in the newspapers of the West, that every man coming down here can have 80 or 200 acres of cotton land, according to his means for working it, to work for one year. Two hundred acres of land means 200 bales of cotton, the net price of which, in New York, will be $40,000. If 80 acres, it will be $16,000. With hired labor, cotton can be raised at 5 cents per pound, which gives a profit of 45 cents per pound net. No farmer of the North ever dreamed of such profit; and if the advantages of coming here were known, they would flock down here by thousands.

"This matter should be brought to the notice of the Government. *I want the man of moderate means*, our Western laborers, here. They will be a militia to take care of the country, and our troops can go elsewhere. *The persons who cultivate the next cotton crop are the ones who will buy the land here.* Shall this land be distributed and owned in small tracts? For the good of the slaves freed by Mr. Lincoln's proclamation I wish it; for to a great extent the ground will be tilled by their labor, and I want a large population of white people here, so their labor will be in demand and respected, and combinations of a few capitalists cannot be made against them. We have uprooted one aristocracy here; let us not establish by our own act one of another kind.

"The question of title to the land must not make timid a man who is thinking to come here. *The cultivation of one year is enough to induce him to come.* A man that takes only 80 acres can go back home at the end of the year with at least $8,000 in his pocket. Would he make one-tenth that by staying at home? See what you can do towards sending the thousands to our gold fields, and locating a large population on the banks of the Mississippi river. Any man who has seen the emigrants going to California and Pike's Peak knows the inducements and recommendations for coming here. Every officer I have spoken to on the subject favors it. Those who want to plant largely, and be the future aristocrats here, oppose it. Now is the time to change the destiny of this country. I hope you will work favorably and immediately for it."

It will not be doubted that, in multitudes of private circles, in parlor, counting-house, exchange, and wherever men do mostly congregate, the question has been constantly debated as to what is to be the condition of the rebel States, and what the attitude of North and South, when this rebellion shall have been crushed. It is alleged that subjugation will be succeeded by a sullen and scowling submission to the laws, a peace in name only—that, under this novel dominion of the laws, the South will chafe and fret, and be as intractable and malignant as ever. A real, hearty, lasting peace, a thorough fraternization, is predicted as impossible. Old associations divided by the sword, are presumed to be beyond the hope of reunion. The nation, nominally compacted, will be in reality an enforced association of radically antagonistic elements, liable to be again convulsed by rebellion, and by this liability so weakened as to become dangerously open to foreign aggression. Constantly on guard against domestic treason, it will be impossible to combine against foreign attack. Intercourse between the sections will be greatly diminished—business between the two will never revive—old friendships will die out—no new ones will be established—and so general an estrangement must occur as to convert us permanently into two distinct and hostile communities.

These are not such views as I either entertain or desire to express. I give them as the utterances of others; and it may be added that, while they are peculiar to one class of thinkers, they are in direct conflict with those of another class, whose habits of thought and action entitle their opinions to be received with equal deference. Between the two, let history and experience decide. The annals of almost every nation are luminous with instruction touching the *finale* of such a crisis as this, because all governments have been subject to similar convulsions. Rebellion seems to be a chronic infirmity of nations. None have escaped it; many have repeatedly experienced it; most of them have survived it. It involves the single certainty that somehow, and at some time, it must come to an end. As we know how other rebellions have ended, we may infer results as likely to succeed the termination of this. I grant that the long smothered but fierce heartburnings which precede and precipitate them, are not, have never been, and cannot be immediately forgotten. In some instances, they have been wholly obliterated in a single generation. In others, they have survived for ages. Scotland has no scowl for England now, notwithstanding the murderous outbreak a century ago, and the equally bloody war of the roses no longer lives in personal animosities; yet Ireland continues sullen and untamable. Even the Reign of Terror survives only in Parisian history. But the desolation of modern Greece remains fresh in the public memory, because the family of Bozzaris still exists.

Our own history, however, furnishes abundant illustrations of how rebellions end. They are of greater significance, too, because occurring among ourselves. What the past has done for us we may feel assured the future will accomplish. The Revolutionary war closed with the withdrawal of the British armies from our shores and the breaking up of the very foundations of the vast Tory society whose opposition had greatly increased the horrors of the contest, as well as prolonged it. These domestic enemies were resolutely refused either protection, indemnity, or even immunity, by the treaty of peace, whose terms were dictated by our commissioners. The public exasperation against them was deep and universal. No act of oblivion was passed by any of the States, but banishment and confiscation was the rule. The leaders became fugitives and beggars. Even the rank and file fared but little better. As many of the former as the British fleets could carry away, sailed with the troops for England. The vast remainder, thus abandoned by England, fled the country from every outlet by which they could escape. The southern Tories sought refuge in Bermuda and the West Indies. Those in the middle and northern States escaped to Canada and Nova Scotia, where they settled in numbers so large as to give to British power in those regions its first successful momentum. The breaking up of families by this terrible exodus occasioned indescribable suffering. Thousands fled from rich homesteads and ample means, carrying with them little else than the clothes upon their backs, losing all they possessed, and being refused permission to return. The less active Tories, the mere sympathizers, who secretly desired the British to succeed, just as the sympathizing traitors of the present day desire the rebellion to triumph, remained in the country. But they lived despised and hated. Honest men shunned them, and society spewed them out. Many, unable to live under an intolerable odium, abandoned locations where they were known as Tories, and sought new homes among strangers. The breaking up of families from this cause occasioned widespread suffering. The numerous gangs of marauding Tories, now represented by the rebel guerrillas, were forced to quit the neighborhoods they had desolated, the people whom they had outraged executing the task with sanguinary thoroughness.

The rebellions which succeeded the Revolution were mere military episodes, though at times presenting an alarming front to the then feeble authorities. Shay's rebellion ended with the dispersion of his followers, the flight, capture, and pardon of its leaders, with an amnesty to the rank and file on returning to their allegiance. The more formidable Whisky Insurrection in Pennsylvania ended quite as ignominiously to its instigators. As in the case of Shay's, that outbreak was fomented by a single demagogue, who, fluent of speech and reckless of results, traversed the country and roused the people to arms. He in turn became a fugitive, with multitudes of his followers. Others were ruined pecuniarily, while two were condemned and sentenced to death for treason, though subsequently pardoned, while a proclamation of amnesty secured a general pacification. The leaders banished, the masses were forgiven. In Dorr's rebellion the same general facts and consequences are prominently visible. But neither of these insurrections developed the murderous hatred which marked the contest between Whig and Tory, or the still more bloodthirsty virus with which the Slaveholders' Rebellion has shocked the world. Yet time, the great pacificator, has obliterated all the bitter feeling which these half-forgotten contests engendered. With the death of the generations that shared in them, it passed away; and the little of it that still survives exists only in the handful of veterans who yet lag superfluous on the stage of life.

Thus the thick-coming future must be conjectured from the clearly defined past. As all rebellions come to an end, so will the present one. The nation will be no more likely to tolerate its bloodstained instigators within its bosom than our fathers tolerated the masses of the less guilty Tories. Though this government has never executed a man for treason, yet it will seem to many that the gallows may now come actively into use. Flight or the halter will be the only alternative for leading traitors whose hands are dripping with loyal blood. The former will expatriate thousands. That secured, the national horizon will be comparatively clear, and amnesty for the masses will develop the germs of a pacification.

But it must be evident that time alone can make it complete and permanent. Our habits as a people, we know, will go far to hasten in the new millenium. Commerce and trade, the modern humanizers, will rapidly reopen every channel now obstructed. Collision of the scales and yardstick will abrade a multitude of asperities, for with too many of us the dollar is the only divinity to be worshipped.

The nations of Europe which, six years ago, were waging war against each other, have all recovered their commercial equilibrium. Southern destitution of a thousand comforts and necessities will compel a grateful recognition of Northern abundance. A vast railroad system, telegraphs, and steamers, created for the promotion of intercourse between the sections, will be actively engaged in facilitating that intercourse. The arrogance of Southern temper will be modified under the subduing knowledge that its chivalry has been soundly whipped by Northern mudsills. Opinion, both printed and spoken, will be free—for Northern colonization will pour in with schools and ploughs, educating a hitherto benighted people, and redeeming an almost ruined agriculture by placing the manufacturer side by side with the producer. This intercourse will rapidly teach the South how grossly her banished demagogues have deceived her as to the design and purpose of the North; and, though humbled by subjugation, she will discover that her prosperity is becoming greater than ever.

But it is idle to presume that the griefs, the passions, the fierce animosities, engendered by this awful contest, will die out while this generation lives. Too many brave men have perished, too many homes have been made desolate, too many families have been broken up and beggared for that. Men whom it has impoverished will live and die poor, remembering constantly the causes of their poverty. Widows will weep over husbands, children over fathers, slain in battle. No catalogue of griefs and horrors could be longer or blacker. These things must be identical with those which closed up the Revolution, only ten times magnified. In those Southern States where loyal men have been outraged by their neighbors, the latter will doubtless be exterminated or driven off by those whom they have persecuted. Blood for blood will be the rule with them. Here, in the North, every local traitor will be marked. Good men will shun him, honest men refuse to trust him, society will keep him at arm's length. His generation will never cease to remember that he was a traitor. The status of these is defined already. Thus history reproduces itself, and we are living witnesses of the instructive truth.

But underlying and overshadowing these general facts, there remains the great question of American slavery. In all former contests, both foreign and domestic, slavery was passive or incidental. In this, it is confessedly supreme, the sole animating cause of rebellion, giving it impulse at the beginning and vitality during its progress. Had it been struck down at the outset, the rebellion would have been brief and far less sanguinary. There has been no pacification during sixty years of its aggressive existence—there can be none so long as it may be continued. It has the nation by the throat, and the nation must have the courage to shake it off and destroy it. I have shown how other rebellions have ended, but it is only by universal emancipation that this one ought to be crushed.

Yet even while the war was being waged, the

South was stretching forth her hands for Northern wares and merchandise. While willing to fight, she was even more anxious to trade. From the first gun at Sumter, every lying trick that treason could devise has been practised to smuggle into the South the products of Northern workshops. A perjury was regarded as commendable that secured the admission of a handful of percussion caps. No false swearing was too black to obtain an ounce of quinine. The dearth of Northern products compelled Southern women to remain at home or go abroad in meaner stuffs than are worn by paupers in a Northern almshouse. Every intercepted letter of a Southern woman called for clothing as a necessity, pins and needles as blessings, and bonnets as the greatest of mercies!

With necessities thus embracing every department of human society, it will be impossible for the South to stand aloof from the North. Agriculture, trade, and commerce, will be greater necessities with her than ever. Her people are now impoverished, and they must begin life anew. As this grasping after Northern products prevailed during the contest wherever there was hope of its being gratified, so will it become stronger and more general with peace. Trade of all descriptions will immediately revive. War has fulfilled its mission—her people have had enough of it, and the old antagonism on the slavery question will no longer disturb what will soon become a general harmony of interests. Commerce will compel a thorough and lasting fraternization. The South will be far safer for Northern emigrants than it has been, during three years, for natives of her soil.

The quantity of land to change owners and be settled up is almost incredible. In Virginia alone, according to Judge Underwood, there are more than two hundred million dollars' worth of property, chiefly real estate, which ought to be confiscated. Thousands of acres have already been sold to Northern purchasers. "These States," says Mr. Julian, "constitute one of the fairest portions of the globe. They are larger in area than all the free States of the North. They have a sea and gulf coast of more than six thousand miles in extent, and are drained by more than fifty navigable rivers, which are never closed to navigation by the rigor of the climate. They have at least as rich a soil as the States of the North, yielding great wealth-producing staples peculiar to them, and two or three crops in the year. They have a finer climate, and their agricultural, manufacturing, and commercial advantages, are decidedly superior. Their geographical position is better, as respects the great commercial centres of the world. The institution of slavery, which has so long cursed these regions by excluding emigration, degrading labor, and impoverishing the soil, will very soon be expelled. The cry which already comes up from these lands is for free laborers. If we offer them free homesteads, and protect their rights, they will come. John Bright, in a recent speech at Birmingham, estimates that within the past year 150,000 people have sailed from England to New York. Let it be settled that slavery is dead, and that the estates of traitors in the South can be had under the provisions of the Homestead law, and foreign immigration will be quadrupled, if not augmented ten fold. Millions in the Old World, hungering and thirsting after the righteousness of free institutions, will flock to the sunny South, and mingle there with the swarms of our own people in the pursuit of new homes under kindlier skies. Immigration has not slackened, even during this war, and in determining the direction it will take, it must be remembered that settlements have very nearly reached their limits in the North and West. Kansas and Nebraska are border States, and must so continue. Their storms, and droughts, and desert plains, give a pretty distinct hint that the emigrant must seek his Eldorado in latitudes further South. In the new northwestern States the richest lands have been purchased, and vast portions of them locked up by speculators. Their distance from the great markets for their produce, and their severe winters, will also check emigration in that direction, and incline it further South, if lands can be procured there with tolerable facility. The rebel States not only abound in cheap and fertile land, with cheap labor in the persons of the freedmen to assist in its cultivation, but they possess great mineral resources. They have also extensive lines of railroads, which, in connection with their great rivers, bring almost every portion of their territory into communication with the sea."

With slavery extinct, and peace restored, then, in the eloquent language of Solicitor Whiting, "the hills and valleys of the South, purified and purged of all the guilt of the past, clothed with a new and richer verdure, will lift up their voices in thanksgiving to the Author of all good, who has granted to them, amidst the agonies of civil war, a new birth and a glorious transfiguration. Then, the people of the North and the people of the South, will again become *one people*, united in interests, in pursuits, in intelligence, in religion, and in patriotic devotion to our common country."

No one who has not visited Virginia since her desolation came upon her, can imagine how complete and terrible it has been. When the army first penetrated the country beyond Alexandria, it was asserted that the corps of axe-men was so large that it levelled an acre of timber every ten minutes. But much of this East Virginia land had been rendered barren by a ruinous system of cultivation before invasion came. Farms were abandoned as worthless, and were sold at one to five dollars per acre. Yet there is no region in the Union containing finer land than this. It is near to Alexandria, Washington, and Georgetown, cash markets in which all that a farmer can produce will sell at high prices. Its soil is capable of the highest improvement at a moderate cost, as was proved by numerous Northern farmers who settled there previous to the rebellion. They made it yield 40 bushels of wheat and 100 of corn to the acre, and raised the market value of their farms from $5 to $60. All these now desolated and abandoned lands must change owners. They can be purchased at extremely low figures.

Their location and advantages are so superior, that of all the rebellious soil of the Southern States, these will probably be first purchased and redeemed by Northern farmers. Such of them as settled there before the rebellion were rapidly becoming rich. The chances for those who may settle there hereafter will be infinitely better.

CHAPTER XIII.

Many kinds of Farmers—Women managing Farms—Very Small Ones—Eleven Acres—A Two-acre Farm—The Spade and the Fork—A Single Acre—Heads better than Hands—Help Your self.

There are all classes of farmers, the sick and the well, the sound and the cripple, women as well as men. Some are cultivating their thousands of acres, using the steam-engine as a ploughman; others are contented on a single acre, depending on the spade and hoe. Yet all seem to live. That they continue to do so is presumptive evidence of intrinsic goodness in the occupation, or that, if a poor one, they manage it so prudently as to make it a paying one. Some of them have been suddenly placed in charge of a farm, with no previous knowledge of the business, yet have done well. Thus some men may be said to be born farmers, as others have been born generals.

It is related by an agricultural journal, that an eminent London tradesman had married the daughter of a farmer who held three hundred acres near London, and who had acquired considerable property. The farmer having died, the widow carried on the farm, but after two or three years' experience discovered that she was losing money. The son-in-law was consulted as to what it was best to do. On looking over the farmer's old accounts, he found that the farm had paid well before his death, and knowing no reason why it should not still do so, under proper management, agreed to take the farm himself. He had been eminent as a gunsmith, and now commenced as an agriculturist. Knowing literally nothing of farming, he began by reading all the books and papers on the subject which fell in his way. He had not read far before he found that a knowledge of chemistry lay at the foundation of good husbandry. He therefore put himself under the tuition of an intelligent working chemist until he made himself a good practical chemist for agricultural purposes. He then applied this knowledge by adapting his manures to the quality of the soil and the nature of the crop he intended to raise from it.

The result was that the neighbors, who began by ridiculing the "cockney farmer," and who prophesied his ruin in three years, were glad, at the end of that period, to go to him for advice about their crops. His own crops of grain, hay, and roots, were the admiration of the whole country, and his wheat would often command more than the market price for seed. At the end of the fourth year, in making up his account, he found a balance of twelve hundred pounds in favor of the farm. Such was the result of science diligently acquired and judiciously applied; and by such means, without any previous knowledge of the subject, did one who had been eminent as a tradesman become equally eminent as an agriculturist.

A lady is now residing in New Jersey whose case is quite as remarkable. She was living with her aged father, on a farm encumbered with mortgages to its full value. He offered to convey it to any one of his children who would agree to keep him for the remainder of his life. All declined the offer except the daughter. Her husband was in feeble health, could be of no assistance, and died soon after. But she engaged resolutely in the work she had undertaken, grasped with surprising readiness the whole details of what, in the eyes of her neighbors, was a hopeless case, and went on prosperously. There happened to be a large quantity of currant and gooseberry bushes on the place. She caused the fruit to be gathered, and converted it into excellent wine, for which she found ready sale in New York. Taking the hint from this, she enlarged her operations another season by buying all the common wild blackberries and currants that were brought to her. The farm was in an isolated location, with no ready sale for perishable fruits; but a market being thus established by her, supply followed demand. All the children for miles around took to picking blackberries, and the quantities offered to her were immense. With characteristic energy she enlarged her facilities for handling them, and bought all that came, converting some into wine, some into syrup, and some into simple preserves.

Meantime, her ordinary farming operations went on with unabated energy. Without doing any of the work herself, she saw that it was done thoroughly and well. In her wine manufacture she was singularly successful. Her products obtained a high reputation, and all that she could make was readily disposed of. The business was profitable, as was shown by her making repeated payments toward freeing her farm from debt. When the great army demand for blackberry syrup sprung up in 1861, she supplied vast quantities at high prices. This demand increased yearly as the war continued, but she enlarged her supply, obtaining higher prices each year, until, in 1864, after seven years of energetic devotion to her farm, she paid off the last dollar of the debt by which it had been encumbered. This woman now owns an unhampered homestead, and is carrying on a highly profitable business, made up of items which the majority of cultivators consider of no commercial value. Having the sagacity to discover their availability, and the skill and energy to develope it, she has had an abundant reward.

There are numerous cases of American women who have been, and who still are equally successful in the management of farms. In England, the devotion of women to agricultural pursuits is even more general and thorough than with us. Mr. Holcomb, in his address before the Maryland State Agricultural Society, relates the following incident as showing the thorough knowledge possessed by some English women on these subjects.

"I cannot but relate a casual interview I chanced to have with an English lady in going up in the train from London to York. Her husband had bought a book at the stand as we were about starting, and remarked to her that it was one of her favorite American authors—Hawthorne. I casually observed that I was pleased to see that young American authors found admirers with English ladies, when the conversation turned on books and authors. But I said to myself pretty soon, 'this is a literary lady—probably her husband is an editor or reviewer, and she uses the scissors for him; at all events, I must retreat from this discussion about authors, modern poets, and poetry. What should a farmer know critically of such things? If I were only in those fields, if the conversation could be made to turn upon crops, or cattle, then I should feel quite at home.'

"I finally pointed out a field of wheat, and remarked that it was very fine. The lady, carefully observing it, said, 'Sir, I think it is too thin—a common fault this season—as the seeding was late. Those drills,' she added, turning to her husband for his confirmation, 'cannot be more than ten inches apart, and you see the ground is not completely covered—twelve, and even fifteen inches, is now preferred for the width of drills, and two bushels of seed to the acre will then entirely cover the ground, on good land, so you can hardly distinguish the drills.'

"If the goddess Ceres had appeared with her sheaf or cornucopia, I could not have been more taken by surprise. A lady descanting on the width of wheat drills, and the quantity of seed! I will try her again—this may be a chance shot, and remarked in reference to a field of ploughed ground we were passing, that it broke up in great lumps, and could hardly be put in good tilth. 'We have much clay land like this,' she replied, 'and formerly it was difficult to cultivate it in a tillage crop, but since the introduction of Crosskill's clod crusher, it will make the most beautiful tilth in these lands, which are now regarded as our best lands for wheat.'

"Conversation turned on cattle—she spoke of the best breeds of cows for the pail, the Ayrshires and the Devons—told me where the best cheese was made—Cheshire; the best butter—Ireland; where the milk-maids were to be found—Wales. 'Oh,' said I, 'I was mistaken. This charming, intelligent woman, acting so natural and unaffected; dressed so neatly and so very plainly, must be a farmer's wife, and what a help-mate he has in her! She is not an extravagant wife, either—not an ornament about her—yes, a single bracelet clasps a fair, rounded arm—that's all.'

"The train stopped at York. No sooner had my travelling companions stepped upon the platform than they were surrounded by half a dozen servants, men and maids, the men in full livery. It turned out to be Sir John and Lady H. This gentleman was one of the largest landed proprietors in Berkshire, aud his lady the daughter of a nobleman, a peeress in her own right; but her title added nothing to her, she was a nobleman without it."

It is not the size of a farm that in all cases determines the question of success. Whether large or small ones are the more desirable, is referred to elsewhere. Great things have been accomplished on the former, but results comparatively marvellous have been achieved upon the latter, sometimes with very inadequate means. No farm is to be despised because it happens to be a very small one. An acre of land possesses capabilities which few can be made to understand until they see them fully developed. It is not exclusively from the land that profit comes, but from the judicious application of labor upon it.

An admirable specimen of farming on a small scale is presented in the management of Mr. Nathan G. Morgan, of Union Springs, New York. That gentleman formerly possessed 300 acres, which he subsequently reduced to 160, and afterward, in consequence of protracted illness in his family, he removed to another place containing only 11 acres. He has remarked that even this is too large. Yet from this little spot he has sold $300 worth of farm products in a single year, besides retaining enough for the use of his family. He performs all the labor with his own hands. He is especially successful in raising pork, and finds this the most profitable branch of farming, much more so than raising wheat. He long since gave up raising cattle, as being far less productive. He has raised 130 bushels of shelled corn per acre, but the average is about 80. By his skill in the art of pork-making, he realizes a dollar per bushel for corn when the pork is five cents per pound in market. He makes an acre of ground maintain a horse during the whole year, by soiling, feeding corn, &c. He thinks, nevertheless, that a large farm may be made as profitable as a small one, if equally well managed; but he considers the temptation, in nearly all cases, is to do the work too superficially.

Coming down to what has actually been done on a two-acre farm, we obtain some approximate idea of the real capabilities of well managed land. A writer in the New England Farmer says, that nine years previous he came into possession of a two-acre farm, from the whole of which it was barely possible to get one ton of hay, so badly had the land been run down. Yet he increased the product of hay to two and one-half tons, and the whole money value of the two acres to $133 per annum. Let only six or ten acres be farmed with equal skill, and any one can cypher up how far the yield will count in the keeping of a family.

The further one looks into this branch of the subject, the more apparent does it become that success depends not on the quantity of land, but on the management. Some years since a little treatise was published in London, by Mr. John Sillett, setting forth the results obtained by cultivation of two acres with the spade and fork. The author being broken down in health by long confinement to business in London, purchased two acres of land, for which he paid $1,180. He undertook farming in total ignorance of the art; yet, he supported himself, wife, and child, entirely from the products of his little tract. After twelve years experience he speaks with confidence of the possibility of a man's getting a living from two acres. He states the requisites to be a fair start with a good piece of land, sufficient means to commence with, skill, perseverance, a willingness to labor, and a reasonable degree of economy.

This success on two acres was secured by rejecting the plough and depending on the spade and fork, the latter being subsequently used as the preferable tool. His land had been many years in pasture, but he produced wheat, potatoes, cabbages, turnips, and mangolds. Of course he owned a cow, which fattened a calf annually to better profit than could have been secured by selling milk or making butter. The litters from a single sow he fattened and sold. The manure he needed was manufactured on the farm, and its efficiency greatly increased by keeping it under cover. While these curious results were being annually realized, Mr. Sillett contrived to build a dwelling-house, cow-house, and piggery, and was contemplating underdraining.

But other benefits accrued from his undertaking. He says—

"Besides the greatest of all benefits that I have derived, in restoring a sickly constitution to perfect health, I felt delighted at the thought of being independent of the harassing cares of business. Of all the feelings which we possess, none is dearer than consciousness of independence; and this no man who earns his living by the favor of the public, can be said to enjoy in an equal degree with the husbandman. In trade there is a great jealousy and competition existing, and a submission to the public which is galling to the spirit. But since I have given my attention to the cultivation of the soil, I find I have no competition to fear, have nothing to apprehend from the success of my neighbor, and owe no thanks for the purchase of my commodities. Possessing on my land all the necessaries of life, I am under no anxiety regarding my daily subsistence."

Spade husbandry means, and in reality is, thorough cultivation of the soil. Our gardens are all perpetual illustrations of its superior value. There is a man at Javington, in England, Dumbrel by name, with a wife and several children, very poor, and so afflicted by disease as to disable him from working a whole day through. An acre of land was let to him on condition of his stall-feeding a cow, building a shed for her, and making a tank for her liquid manure. At first his neighbors ridiculed him for keeping his cow under shelter all the year round, saying it would be unhealthy. But his answer was that some one had lent him £5 toward buying a cow on these conditions, and he would try them. Eminent agriculturists, hearing of this poor man's case, went to see how he was succeeding with his cow. They found her perfectly healthy, after being stall-fed three years and a half, and that from the butter he had sold he had soon paid off the loan of £5, and had got a second cow on half an acre of pasture, but he told them he had had to apply twice to the farrier for the pastured cow, but never for that which was stall-fed, while she gave a third more butter than the other. This land was worked with the spade, and thus was made to yield a succession of green and succulent crops for summer use, with abundant store for winter consumption, besides supporting a large family. Dumbrel's condition was improving annually. Yet in his case there was a combination of two exceedingly discouraging elements—ill-health, and spade-husbandry, the most laborious description of agricultural toil. Its superior value was manifest in making a sick man entirely comfortable.

More remarkable than any of these cases is that of a farmer who rose from nothing into absolute independence, though born without either hands or arms. This man was William Kingston, the son of a laboring woman, and was ushered into this world poor and helpless, for he had neither hands, arms, nor shoulders. But nature had blessed him with longer toes, and greater flexibility of feet and legs, which by constant use made up for the deficiency of hands and arms. He shaved himself regularly, wrote plainly and distinctly, and in dressing and undressing required assistance in buttoning and unbuttoning only certain portions of his dress. He was at no loss at meals; tea, coffee, and food were conveyed by his feet with equal facility as by hand. In the hayfield he was as active as any other in securing the crop, and performed every duty in the haymaking process, except mowing and pitching. Few were better milkers—he worked at all the requirements of a dairy-farm, with the aforesaid exceptions. He could cut the hay at the stack, and take it to the cows and fodder them as well as others did.

But this deficient body was furnished with a capacious head in which a mighty brain was stored. When a mere boy, some compassionate neighbor gave him a hen and chicken, then another gave him a lamb. These fractions of a capital he nursed and multiplied until he was able to procure a colt. In the end he became possessed of a dairy-farm, and died independent. It would seem clear that his success depended exclusively on his head, seeing that he was destitute of arms.

Few facts could show more conclusively than this, that he who is striving for success in life must be his own right hand-man. A writer who is unknown to me, lays it down as a rule that "people who have been bolstered up and levered all their lives, are seldom good for any thing in a crisis. When misfortune comes, they look around for somebody to cling to or lean upon. If the prop is not there, down they go. Once there, they are as helpless as capsized turtles, or unhorsed men in armor, and they cannot find their feet again without assistance. Such silken fellows no more resemble self-made men, who have fought their way to position, making difficulties their stepping stones, and deriving determination from defeat, than vines resemble oaks, or sputtering rushlights the stars of heaven. Efforts persisted to achievements train a man to self-reliance; and when he has proven to the world that he can trust himself, the world will trust him. It must therefore be unwise to deprive young men of the advantages which result from energetic action, by buoying them over obstacles which they ought to surmount alone. No one ever swam well who placed his confidence in a cork jacket; and if, when breasting the sea of life, we cannot buoy ourselves up, and try to force ourselves ahead by dint of our own energies, we must go to the bottom."

"We must all learn," he adds, "to conquer circumstances, thus becoming independent of fortune. The men of athletic minds, who left their marks on the age in which they lived, were all trained in a rough school. They did not mount into their high positions by the help of leverage. They leaped into chasms, grappled with the opposing rocks, avoided avalanches, and when the goal was reached, felt that but for the toil that strengthened them as they strove, it could never have been attained."

The question whether it is better to have a small, well-cultivated farm, or a large one poorly cultivated, or not cultivated at all, and whether men with small holdings are not usually most successful, has often been discussed. It would be out of place to reopen the debate here. But as the subject of small farms is now before us, I quote from the *Country Gentleman* the following remarks of Mr. O. S. Leavitt, of Detroit, as bearing on the question—

"The general broad proposition, without qualification, that small farms are preferable to large ones, is a fallacy. A great deal of nonsense has been written and spoken in favor of it, and, in my judgment, no one error of opinion is more wide-spread or more injurious. The real arguments, *pro* and *con*, lie within a very narrow compass. All will admit that it requires a great deal of science, knowledge, industry, tact, and agricultural skill, to conduct even a farm of 20 acres in the best manner, or in an excellent manner. Very few farmers can do it. The cases are rare, indeed, where a small farm or a large one either, is well managed. I am not writing to flatter the farmers; I would rather tell the truth. Are all the small farms well managed? Far from it. Probably more large farms, in proportion, are well managed than small ones. If a farmer is to do all his work himself, he of course should have a small farm; but it is absurd to say that a man who thoroughly understands farming, say one farmer out of a hundred, should do this. A *good* farmer would greatly benefit the public, as well as himself, by hiring as many unskilled farmers as he can superintend, and let his light shine. In fact, so few farmers understand their business, it would be better for them generally to hire out to some good farmer until they can learn how better to conduct farming on their own account.

"Now, it requires more skill to conduct a small farm profitably, than a large one. A greater variety of products is required, all of which must be understood; more care is required, greater judgment is necessary, and even the very drudgery must be mostly done by the owner himself. A farmer of less agricultural skill, confining himself to such branches as he best understands, with only ordinary judgment, hiring judiciously, &c., will generally do better on a large farm than a small one. But there are so many contingencies and conditions that may affect the result, it is very difficult to lay down a proposition on the subject sufficiently clear and guarded to be of general application.

"Then this is a day of agricultural machinery; this, indeed, constitutes a new epoch in the annals of agriculture. The mowing and harvesting machines are of yesterday. Can the twenty-acre farmer avail himself of these? Can he even supply himself with all the cheaper and smaller improvements—ditching machines, planters, drills, rollers, horse hoes, clod crushers, potato diggers, horse powers, threshers, &c.? Nor can he generally avail himself as well as the larger farmer, of those natural resources, often so useful and so neglected, of running streams for irrigation, supplying water for stock, &c., to say nothing of his greatly increased expense for fences.

"The constant reader of the *Country Gentleman* is aware that successful agriculture requires a wider range of knowledge, more extensive and varied attainments, than almost any other pursuit. Indeed, but few men can succeed well in all its departments. The good dairyman or grazier may be but an indifferent fruit grower; a successful cultivator of the cereals is seldom equally successful at sheep husbandry or horticulture. The man who has been blessed with such varied endowments, and is a genius of so high an order as to succeed well in all these things, should indeed have a large farm, giving employment to as many young men as possible—a model farm—a normal-school farm. Such a man should dispense wisdom and knowledge daily, both by word and action, theory and practice, to great numbers of his dependents, assistants, employés—overlooking, superintending, directing everything—a Columella in wisdom, a Liebig in science, a Mechi in vigor and enterprise. We see him, by the aid of his manager, gardener, nurseryman, and fruit grower, architect and builder, machinist and blacksmith, civil engineer, chaplain, professor of chemistry, schoolmaster, farrier, surgeon, landscape gardener, &c., practically educating his numerous laborers, so that they may in due time be able to carry much of this weight of wisdom into practical effect, and butter their bread upon the countless millions of uncultivated lands now at the West.

"But this may be called a mere fancy sketch—an extravaganza. I shall be told that, in this country, no man can succeed by thus attempting to carry on farming by employing so many skilled, educated persons, as it would be desirable he should do, at the high salaries they would require, and that it would be necessary to have the wealth of a Nabob. Well, if the farmer, having the requisite amount of land, hasn't the means to hire them, *let him take them into partnership.* If he has but the land, and is the true country gentleman of education, experience, business talent and tact, he can easily draw around him the landless, skilled, educated, and industrious persons, to take charge of the various industrial departments of his estate for a reasonable share of interest, and divide the profits among the members. They could then, among themselves, have a school good enough to render it entirely unnecessary to send their children abroad to be educated."

CHAPTER XIV.

Why Land so often changes Owners—Tenures and Estates in England—Absorption there and here—Results of English Husbandry—The real Value of Land—Stick to the Farm—Scarecrows—Why Farming is Unprofitable—Go where most wanted.

It has been assumed, throughout these pages, that the masses in this country are desirous of becoming owners of land. But among the curiosities of the subject is an extraordinary propensity among a portion of them to get rid of it. This must have its origin in the absorbing passion of Americans to become traders and speculators rather than farmers. Some writer, whose name is unknown to me, pronounces the American a type of a restless, adventurous, onward-going race of people. "He sends his merchandise all over the earth; stocks every market; makes wants that he may supply them; covers New Zealand with Southern cotton woven in Northern looms; sends clerks of stores to the Sandwich Islands; swaps with the Fejee cannibals; sends his whale-ships among the polar icebergs, or to cruise in other solitary seas, till the log-book tells the tedious sameness of years, during which boys become men; gives to the torrid zone the ice of our Northern winters; piles up the crystal squares of

Fresh Pond on the banks of the Hoogly; gladdens the sultry savannahs of the lazy South, and makes life tolerable in the bungalow of an Indian jungle. The lakes of New England thus awake to life by the rivers of the East, and the antipodes of the earth come in contact at this meeting of the waters. The white canvas of his ships is seen in every nook of the ocean. Scarcely has the slightest information come of some unknown and obscure corner of the sea, when captains are consulting their charts, and cargoes are taken in for speculation." An idea successfully inaugurated here, he reproduces wherever civilization is established. He covers the West India Islands with a net-work of railroads, astonishes England with street conveniences of the same description, raises fleets of sunken frigates in the harbor of Sevastopol, scents out new guano islands, and gluts all Europe with mowers and reapers, Yankee clocks, sewing machines, and baby-jumpers. It is money he is seeking to acquire—anything but houses or lands.

The reverse of all this is witnessed in Europe. "There," says another, "but more especially in England, landed property seldom changes hands. This is partly in consequence of entails. But apart from that, if a man owns a house or lot, in a town or village, it will commonly remain in the hands of children and children's children, while here a man will buy and sell the homestead a dozen times, or at his death it passes into the hands of strangers, almost as a matter of course. Why is this? A multitude of causes operate to produce it, such as the equal distribution of property among children, the prejudice in many quarters against making a will, the rise and fall of fortunes, and the constant changes in the value of real estate. But one cause is probably more potential than all. Almost every man who buys real estate, buys, builds, or improves, not so much with a simple view to his own comfort, as with an eye to what it will sell for when he has done making his improvements. Many men build very extensively, calculating upon this and this alone. They are not architects, not builders, not even capitalists; but they build largely on credit, simply to wait as tenants for a rise in price, and then sell. Of course, the larger and more expensive the house, the larger the hope and expectation of gain.

"All such," he adds, "are mere speculators. Their plans succeed often enough to encourage others who fancy they have skill and taste to imitate. Many of these aspirants for fortune lose thousands, but their disasters are soon forgotten. Such men, says some one, build houses far in advance of their real means and situation in life, and cities improve and towns spring up, beautiful to look upon, but too often they are not the abode of content and happiness. They might be, if each man built an humble home, just such as he could afford to hold through all the vicissitudes of business—such a home as his widow could afford to own, free from debt, and live in, without pecuniary care, after he was gone—such a home as any one of his children might reasonably hope to be able to keep up, without inconvenience or ostentation. We all see how families scatter; but this is one chief cause of the scattering —that men buy houses, or build them about as a soldier builds his tent, for a night. Houses, farms, and furniture, are bought on credit, kept on mortgage, and sold at a loss or gain, as the case may be, at five minutes' notice, because it was a part of the purchaser's original plan to hold the property only for an advantageous opportunity of disposing of it."

The same writer thinks that, in "nine cases out of ten, it is cheaper to buy than to build, because dealing in houses and lands requires a skill and a capital which the majority of men do not possess. It is, moreover, far more comfortable, because it is free from the cares and annoyances of being engaged in matters very complicated, aside from a man's regular business, and because he can generally select such a residence as suits his wants, and means, and convenience. In Europe, the happiest homes are those which, having been built for the present convenience of the father, have been merely altered so as to make them correspond with the present condition of the son or grandson, or more remote descendant, as the case may be—greatly changed by time, and mellowed and softened by its graver tints, or decayed and nibbled into by its old tooth, or enlarged piece by piece, and generation by generation, irregularly it may be, but with a strict eye to comfort and convenience, and corresponding with the growing wealth of an accumulating and prudent family. In this country, we live too fast in the first generation to have many such houses as these—homesteads built comfortably within a man's income, in which he can and does live more liberally than he cares to appear to live—houses in which, if times should change, he or his children could live consistently and without meanness, on half his present expenditure.

"Another great reason why houses change hands so often, and are so little homes for those who possess them, is the practice of living too hopefully into the future, and without regard to future exigencies, living with too much sail in proportion to ballast. It is for every man to think what he desires in this matter. Many do not desire homes in the sense that I have described. They prefer to live on the wing, and in a splendor above their real means. With such it is better to rent, and quite absurd to build. It is even worse—it is mocking their children to build a house so splendid that none of them will be able to keep it up. Better the humblest cabin, respected and loved as the dear old home of childhood, the shelter of a father's gray hairs, the centre of a hundred happy gatherings round a mother's knees, the refuge uninvaded by all enemies save the last, and where even he is so serenely met and welcomed as to be robbed of his sting, and to have his enmity destroyed."

In England, the passion for owning land may be said to be hereditary among all classes. With the titled and rich, who are able to gratify it, it has produced some remarkable results. The great landholders are now comparatively few in number. I have seen them variously computed at from 30,000 to 40,000, who hold landed property yielding an annual rent of not less than $500, the number rapidly diminishing as the annual rent increases. The incomes of the wealthiest range from $100,000 to $1,500,000 per annum. A hundred years ago the landholders of England were numbered at 230,000, which number has ever since been rapidly diminishing by the purchase of the lands belonging to the thriftless and wasteful, by the more prudent and wealthy.

The Marquis of Bredalbane rides from his house a hundred miles in a straight line to the sea, on his own property. The Duke of Sutherland owns the entire county of that name, stretching across Scotland from sea to sea. The Duke of Devonshire, besides his other estates, owns 96,000 acres in the county of Derby. The Duke of Richmond has 40,000 acres at Goodwood, and 300,000 at Gordon Castle. The Duke of Norfolk's park in Sussex, is fifteen miles in circuit. An agriculturist recently bought the island of Lewis, in the Hebrides, containing 500,000 acres. These vast domains are constantly growing larger. The great estates are absorbing the smaller freeholds as opportunity offers. The great landholders never sell.

Wherever slavery has cursed the soil of this country by its presence, the same process of absorption has been going on. Mr. Clay, of Alabama, gives the following gloomy picture—

"I can show you, with sorrow, in the older portions of Alabama, and in my native county of Madison, the sad memorials of the artless and exhausting culture of cotton. Our small planters, after taking the cream off their lands, unable to restore them by rest, manures, or otherwise, are going further West and South, in search of other virgin lands, which they may and will despoil and impoverish in like manner. Our wealthier planters, with greater means and no more skill, are buying out their poorer neighbors, extending their plantations, and adding to their slave force. The wealthy few, who are able to live on smaller profits, and to give their blasted fields some rest, are thus pushing off the many who are merely independent. Of the $20,000,000 annually realized from the sales of the cotton crop of Alabama, nearly all not expended in supporting the producers is re-invested in land and negroes. Thus the white population has decreased, and the slave increased almost *pari passu* in several counties of our State. In 1825, Madison County cast about 3,000 votes; now, she cannot cast exceeding 2,300. In traversing that county one will discover numerous farm houses, once the abode of intelligent and industrious freemen, now occupied by slaves, or tenantless, deserted, and dilapidated; he will observe fields once fertile, now unfenced, abandoned, and covered with those evil harbingers, foxtail and broomsedge; he will see the moss growing on the mouldering walls, of once thrifty villages, and will find one only master grasps the whole domain that once furnished happy homes for a dozen white families. Indeed, a country in its infancy, where fifty years ago scarce a forest tree had been felled by the axe of the pioneer, is already exhibiting the painful signs of senility and decay apparent in Virginia and the Carolinas."

Among English owners "are many men of the highest intellectual powers and attainments, of the highest social position, and of the most refined culture; noblemen not only by right of geniture and rank, but noble men in the noblest sense of the word, who are carrying forward on their enormous estates the most magnificent operations in the highest culture of the soil, winning from their well fed and well tilled acres the richest reward of the wisest husbandry. One contemplates with amazement the magnificence of their arrangements for irrigating hundreds of acres, as may be seen on the estates of the Duke of Portland; the vast extent of their systems of drainage and subsoiling, the enormous capital invested in carrying on their agricultural processes and improvements, and the vast revenues by which they are enabled to push forward their splendid designs. It is fortunate for mankind that such men devote their talents and their great resources to an enterprise so important, and exert their powerful influence in the promotion of so great a cause —a cause which holds concentrated within itself every inducement to allure the loftiests minds and the fullest means to its support, as upon its success humanity itself depends for the continuance of its very existence."

It may be safely asserted that "but for the high culture which the soil of England has received under such influences, and the consequent development of its exuberant riches, her population could not have made the great strides that have carried it from four and a half millions in 1600 to more than twenty-five millions in 1862; nor could the nation itself have attained that immense power and wealth which make her now to stand foremost among the nations of the world, and the richest of human aristocracies."

Thus, "under the influence of the culture created by the action of such minds upon labor, we find a yield of 50 to 80 bushels of wheat per acre in England, from 40 to 70 in France, and the productive power of an acre of land in the well cultivated part of Europe to be double what it was seventy-five years ago. In proof of the influence of improved tillage in England in enabling her to sustain her own people with diminished reliance on importations from foreign countries, I may here state the important fact, that while in the first ten years of the present century she imported foreign wheats at the rate of eight quarts per annum for each person in the realm, in the next ten years she imported but six, in the next five years but four, and in the last three years of these five, at the low rate of a single pint, the soil of the kingdom supplying all the rest consumed. More land had indeed been brought under tillage, but every acre, old and new, having been better tilled, had made a better yield."

Many of the foregoing facts and observations are here reproduced from anonymous sources; but it may be said that everybody who has taken the smallest pains to ascertain the facts, knows and testifies that agriculture has advanced just in proportion as mind, mind as developed in men of intellect, intelligence, education, and reflection, has given attention to it. The condition of English agriculture, as an obvious and suggestive example, bears ample testimony to the influence of mind upon progress.

When Americans acquire a similar fondness for agriculture, the possession of land will be as eagerly desired by them. I admit that such a passion has sensibly increased within twenty years; but our country is too vast in size, our population too limited in numbers, and the wealth of the community too small, for the present century to realize any approximation to the intensity with which it animates the popular heart in Europe.

The Philadelphia *Ledger* says:

"It has been remarked, by political economists, that the sum per acre at which landed property is found to sell in nearly all countries, and for long periods, proves that men are glad to purchase it at a price at which they know beforehand it will yield them less than the ordinary rate of interest. In this country, men who could, on bond and mortgage, easily get six or seven per cent., will be glad rather to purchase the same lot of land, even though they can hardly thus make five; and in England, where ordinary interest is five per cent., they will purchase land where it will not yield more than three.

"Various have been the methods taken to account for this well-established fact. Some have attributed it to the tendency of landed property slowly and steadily to rise in value as population becomes more dense, or roads and schools become better, and the arts and habits of civilization more complete. This may in part account for it, especially as money has a tendency to depreciate in value very noticeably in a long period. But after making every calculable allowance for that, the whole of this tendency in man to become possessed of land at a less remunerative price is not fully accounted for. The superior supposed security of landed property has something to do with it. A man who has a mortgage may find some fraudulent evasion of it, some sale for taxes which has destroyed his title. Or by the death of the proprietor lawsuits may be brought that will nearly swallow up the value of the mortgage in securing the claim, so that it has become an invariable rule among capitalists *not* to invest even on mortgage where the property is so far off that the man interested cannot look after it himself. But land, if you start with a clear title, and pay the taxes, and are loyal and quiet, is safe, and rises with all the improvement of society, to fall only with its decay.

"But the real and most fundamental reason for the extra value attached to land, it appears to us, is, that however narrow its dimensions, the possession of every acre involves rights, the extent of which is literally unbounded. Stakes may be driven at the four corners of a lot, and the lines marked out and fenced in, north, south, east, and west. But there are at least *two sides* on which it has *no bounds*, the height above and the depth below. No one knows the values that are or may be connected with those unlimited sides of every man's farm. 'Build high,' was a saying of Stephen Girard, 'there is no ground rent above.' No one knows what commercial value may some day attach to a good situation for drawing in unlimited supplies of oxygen or nitrogen. The fresh air or the beautiful prospect even now give their greatest real value to many a building-lot of land, for the eye has an empire of its own. It can stretch far and wide, and from the proper stand-point connects the possessor of land with realms and a domain stretching out beyond the farthest star into absolute infinity. Some day we may navigate the air, at least the lower regions of it, and what may be the value of those rights, now only used by the tree tops and the birds, who shall say? The shooting is even now of great pecuniary value in England.

"But it is not only above, but below that most appreciable value is likely to become increasingly connected with the possession of land. All the wealth of minerals and mines is thus obtained. To say, however, nothing of these possibilities, how deep we may finally penetrate into the bowels of the earth for water, or what may springs furnishing salt or oil afford, who shall say? Already we suck, if not honey out of the rock, at least '*oil* out of the flinty rock.' We may before long tap the earth regularly for gas or fuel, or still more directly for heat or motive power.

"There is, however, yet a third direction in which the possession of land connects a family with the unlimited—it is theirs, till voluntarily parted with, *forever*. It is like a spring; it yields a present supply, and one that is undiminishing forever. The farms of England, that have been worked for ages, yield on an average far more and not less to the acre to-day than those of the State of New York or Pennsylvania, that have been cleared but half a century. The supply is perpetual. While the earth remaineth, seed time and harvest shall not cease.

"Every spot of ground thus connects its possession with the limitless on *three* sides, and it is this sublime alliance of the mind of man with the unbounded and the infinite, which gives to the possession of land such a peculiar value. It enlarges the ideas more than any other earthly possession, and leads out the soul into all that is grand and sublime."

After all that has been said touching the best way to get a farm, and where to find one, it may be added that almost everything depends upon the man himself—he must either know how, or he must learn how to manage it. It has been well said, that tools do not make the workman, but the trained skill and perseverance of the operator himself. It is proverbial that bad workmen never yet had a good tool. Some one asked Opie by what wonderful process he mixed his colors. "I mix them with brains, sir," was his reply.

Having secured possession of a farm, let the young man cling to it. Now, when land and its products have been daily rising in price, with butter at half a dollar, and hay at twenty-five dollars per ton, be not led away from this honorable and independent occupation by the seductive glitter of some apparently golden prospect in the distance—some chance in city life, where there may be less of bodily toil, more of what the world calls honor, but less of real comfort. On this subject Mr. Henry H. French gives us an instructive caution:

"You are tempted to exchange the hard work of the farm, to become a clerk in a city shop, to put off your heavy boots and frock, and be a gentleman, behind the counter! You, by birth and education, intended for an upright, independent, manly citizen, to call no man master, and to be no man's servant, would become, at first, the errand boy of the shop, to fetch and carry like a spaniel, then the salesman, to fill the place which, at best, a girl would fill much better—to bow, and smile, and cringe, and flatter—to attend upon the wishes of every painted and padded form of humanity—to humbly suggest to rakes and harlots, as well as to starched and ruffled respectability, what color and fabric best becomes the form and complexion of each—and finally to become a trader, a worshipper of mammon, compelled to look anxiously at prices current of cotton and railroad stocks, in order to learn each morning whether you are bankrupt or not, and in the end to *fail*, and compromise with your creditors and your conscience, and sigh for your native hills.

"Or, perhaps, your party being in power, you would obtain a clerkship at Washington, and remove your little family from the North, to a more genial climate, to live at your ease, and grow rich on twelve hundred dollars a year! You give up your little farm, your New England privileges of schools and churches, your independent and influential membership of parish, and district, and town, and church, the woods and playgrounds for your children, your friends, and kindred, and *home*. Twelve hundred dollars is a large sum to you, half the price of your farm, perhaps, twice the amount of the minister's salary. With your habits of economy and thrift, you can live on half the amount. Your arrangements are to be made—the homestead is sold, and you are *landless*. After all, it is not so easy parting with our household gods. The trees our hands have planted take root in our hearts, the vines and roses, twined by our own fingers, and those of our loved ones, over rustic arbors, cling round us more closely than we thought. Your labor has been mingled with the soil of every field. Tears are in the eyes of your wife at every thought of departing, but she trusts in your superior judgment, and no murmur escapes her lips at your decision.

"You have left your home. At the end of a single year, in the city of magnificent distances, you have bitter realizations of the meaning of that phrase. It has proved, indeed, to be full of magnificent distances for you, from happiness, from independence, from advantages of every kind. For the first time, you have felt how sore a thing it is for a Northern freeman to be dependent, to labor at stated hours, at the bidding of a superior officer, to feel that the office you fill, on which depends your very means of living for yourself and family, is held at the arbitrary will of another, who may, if he please, make a servile conformity of your views with his own, on political, or what you may deem moral questions, the condition by which you retain your place. You who, at home, had never seen the man who dared claim to be your superior, are forced to submit to the iron rule of caste, to send your card to the secretary, whom you once knew, perhaps, as an equal, and wait an hour with the colored servant in the hall, to be told at last to call another day—to be slipped over, or shaken off by the 'member' whom you helped to elect, and who now had no further use for you, and consume your energies in endeavoring to keep the toe of your boot from proximity with that part of his person where his honor holds its seat—to be assessed to support party presses whose principles you may despise. In short, you have sold your birthright for a mess of pottage.

"But the half is not yet told, for even the mess of pottage is not sufficient for your wants. Your salary is at starvation point. You must pay two hundred dollars for a house, with two parlors, and a basement for servants, without a cellar, without a closet, without a pump or aqueduct, without a sink, or clothes yard, or garden. Your wife, with the aid of a servant, cannot do the work so easily as she did it alone when at the North. All the water comes in buckets from the city pump a dozen rods off; the slops are poured into the street, your clothing is crammed into wardrobes, your supplies must be procured daily at market, in contemptible quantities—in short, everything, except the parlors, which are for show, and to make you seem respectable, must be richly carpeted and curtained, everything is adapted to the idea that labor is degrading, and that the comfort and convenience of those who perform it is not worth consulting. The thrift, the energy, and comfort of Northern households is unknown in this latitude.*

"Look now at the prices of necessary articles of food. On your farm, however small, your cellar was always filled with an unlimited supply of such vegetables as you desired, and barrels of beef and pork of your own slaughtering. Your granary had always as much of corn and rye, and perhaps of wheat, as you chose to use. Your cows at all times gave you milk and butter in abundance, and your garden and orchard supplied fruits for yourself and the children without stint. Now you buy a peck of potatoes for three shillings, beef at sixteen cents a pound, turkeys at from a dollar and a quarter to two dollars each, chickens with the shells scarcely off their heads, not larger than robins, at twenty-five cents each, butter at thirty-one cents, and milk at eight. Instead of enjoying the abundance of the earth, as you have been accustomed to do, you begin to associate the idea of dollars and cents with the food on your table; you are compelled to vex yourself with economizing in the details of living, and to feel your soul gradually narrowing in, to a conformity with narrow circumstances. You find yourself a poorer man than while upon your hard Northern farm, poorer in your animal means of living, poorer in comparison with those around you, poorer in independence, in prospects for yourself and family, poorer in everything."

This leaving the farm for a fortune is deplored in graphic terms by the author, "Doesticks," who avers that

---

* This was written in 1854. Ten years have revolutionized, redeemed, and purified Washington.

"Young men in the West, when they get too lazy to plough, drive oxen, and dig potatoes, invariably either go to study law, physic, or divinity, or emigrate to New York to make their fortunes. Hence the inundation of two-and-sixpenny pettifoggers, the abundant crop of innocent, juvenile-looking M. D.s, and the army of weak-eyed preachers, whose original simplicity is too deep-rooted to be ever overgrown by the cares of after life. The portion of our country known as the West, sends forth every year scores of these misguided innocents, who, had they stayed at home, might have grown up into tolerable farmers, or even been cultivated into respectable mechanics, but who, being thrown into the whirl of city life, degenerate into puny clerks, with not half enough salary to pay for their patent-leather boots.

"It is a curious fact, that two-thirds of the young men from the country, their first year in the metropolis, do not receive as a remuneration for their valuable services, a sum sufficient to keep them in theatre tickets. If a committee of their employers should be detailed to investigate the hidden pecuniary fountain whence these young men obtain the funds many of them lavish so freely, the said committee would be considerably astonished to find out how much more champagne and oysters the New York merchants pay for than the most knowing of them are aware of; and their wives would be astounded to learn how many bracelets and diamond pins had been presented to ladies of the theatre and ballet, and bought with their husbands' money. And many a country mother would mourn to hear that her darling had, in the first six months of his city life, learned to practice more vices than she had ever heard of, and among his other attainments, had acquired the elegant city accomplishment of spending his employer's money as freely as if it were his own."

I grant that change of occupation is sometimes desirable and prudent, sometimes even necessary. When driven to it by pecuniary calamity, it should be made courageously, and with utter contempt of the jibes or sneers of others. I know the world is full of scarecrows—but why is it that they are always black? Go into the newly planted cornfields, and you will see the unfolding blade protected by effigies as woeful as if clad in the cast off inexpressibles of a rebel regiment. Old hats and older coats, mounted painfully on crooked sticks, or waving to and fro in the wind, suspended from patriarchal apple-trees,—these are scarecrows—but what simpletons the crows must be for being thus easily frightened off from a living. Yet one need not go so far as even the nearest cornfield for a counterpart. There is a scarecrow of some kind in every house, domesticated at every fireside, nestling in every bosom. Old hats and worn out inexpressibles may serve to frighten crows, and we may smile at their simplicity in thus being kept at bay; but who among us, no matter how straitened or depressed, that does not find the prospect of reassuming even his own cast off toggery to be the unsuspected scarecrow of his own heart? Pride, personal, foolish, utterly groundless, is the great national scarecrow. Too many of us are slaves to appearances, mere walking advertisements of the milliner or tailor. Strangely enough, one leaves his brown stone mansion and takes to boarding, without compunction; but the thought of earning bread by apparently mean employment, is absolutely killing, though it be honester than any previous one. This reveals the household scarecrow, for wife and family protest—it were better to starve! We ridicule the crow for his simplicity, but are we any wiser?

I know how habit clutches at the heart—how gambling, rum, tobacco, both rule and stultify—and that this pride tyrannizes with a despotism infinitely more galling, because the whole family become its victims. Yet no heroism can be greater, no good sense more positive, than that which so strengtheus a man into breaking its bonds, grappling with the emergency, be it what it may, and kicking his own particular scarecrow to the dogs.

"Right Side Up," traced legibly on a shingle which had been planed smooth to enable the artist to display his skill, met my eye one morning at a crowded steamboat landing in New York, soon after the rebellion burst upon the country. It stood upon a table covered by a clean white cloth, whereon was bounteous store of peanuts, gingerbread, and pretzels, intermixed with oranges and penny confectioneries for which all urchins seem to have been born with an undying relish. As the day was windy and the street alive with carriages and other vehicles, these delicacies were frosted with dust. Tubs containing bottles of root beer stood shadily under the table, but the imprisoned effervescence must have been hot in spite of the water that surrounded them. These things I barely noticed—it was the shingle that attracted my attention. Could this be one of the signs of the times?

It seemed to mean something, hence I paused to look and consider, never noticing the man who rejoiced in being proprietor of this diversified exhibition. But though not observing him, it seems that he had noticed me.

"Right side up, sir, and mean to keep so," he exclaimed, rising from his seat, coming to where I stood, and extending his hand. My astonishment was really very great, but he exhibited none. His emotion appeared to be that of satisfaction, open, manly, heart-felt. I had known him as the expert foreman of a great machine shop, overlooking several hundred hands, and making twenty dollars a week; but at a glance I conjectured how the matter stood.

"Why, how is this?" I asked.

"Swallowed my pride three months ago," he responded. "Work was dead, employers broke, nothing to do, had a wife and family, only fifty dollars in pocket, and with twenty of it set up this table. It sounds like coming down, and it *was* hard to undergo this publicity, but look at my hands, they are clean, and I am not ashamed to look any man in the face. My pride has left me—gone to trouble some one else. Wife and daughters make the cakes and beer, keep up the table, and the table keeps up us. How does it strike an old acquaintance like you, or have you any thing better on hand?"

Here it all was in a nutshell, just as I had somehow conjectured. I looked upon the man as a hero—he *was* a hero. Depend upon it that all such were not crowded into Pickens or Sumter; for there is a moral heroism which as effectually challenges the admiration of good men as that which personal daring wins at the cannon's mouth. Did he not confess to the scarecrow, and had he not ousted it? Here was evidence of the truly independent, self-relying grit, which, in some form or other, ought to be exhibited in time of sore calamity.

"You have done well," I answered. "You have made a double conquest—pride and bad luck. But this will lead you to better things. There are many stepping-stones to fortune, and this may be yours. I honor you for your courage."

Poor, brave fellow! These must have been among the few words of cheering sympathy he had heard, for his lip quivered, his eye became moist, and he did not trust himself to reply. His heart was bigger than I had ever dreamed of.

The possession of what one may really call a home, is among the unspeakable blessings of this life. The man who cannot prize it must have some radical deficiency in his organization. Mr. Beecher refers to such a possession in one of his characteristic paragraphs. He says—

"But what a poor, shivering, restless, rapping sprite is without a body, that is a living man without A HOUSE. He cannot take root. A man at a hotel is like a grapevine in a flower-pot, movable, carried round from place to place docked at the root and short at the top! There is nowhere that a man can get real root-room, and spread out his branches till they touch the morning and the evening, but *in his own house.* If I could, I should be glad to live in the house that my ancestors had lived in from the days of the flood. That cannot be; for in ascending the line of ancestry I find the people but not the houses, and it is more than suspected that some of them never owned one! My father's house! It is like a picture rubbed out. The frame and canvas are there, but strangers have possessed it. The room where I was born, where my mother rocked my cradle and sang as angels do—where she died, where all my boyish frolics began and life spread out its golden dream—they are all overlaid by other histories. We planted pleasant things in the old house, but the Assyrians came in and settled down upon them."

But some one has said that nine-tenths of our farmers find farming unprofitable—that is, paying but a very small percentage on the capital invested in land, stock, tools, &c. On this subject the editor of the *Country Gentleman* has crowded a mass of interesting information into a nutshell, much of which I reproduce as appropriate to the discussion. He says that hundreds of farmers who own from a hundred and fifty to three hundred acres of good land, tolerably well stocked, find themselves barely able to prove that they are as well off to-day as they were a year ago; and many declare that the laborer, who has nothing but his hands to get a living, lays up more money in a year than they with all their broad acres and flocks of cattle and sheep. If this be true, and doubtless it is so in many instances, a farm managed as a large share of our farms are managed, would be a clog to a young man with a small family, who is endeavoring to lay up something for those rainy days which are sure to fall to the lot of many of us in our journey through life.

It is not high-priced labor, nor unpropitious seasons, nor because produce brings a low price, that makes farming unprofitable. Every laborer is worthy of his hire; the harvests are bountiful; consumers are increasing so rapidly as to keep provisions up to prices that ought to be satisfactory to producers; yet the question as to why farming is unprofitable continues to be asked. It is just as regularly answered, but does not seem to stay so.

Instances are continually reported of men who have maintained families on the product of ten, fifteen, or twenty acres of land, which, when they began, was no better than the average. They have lived respectably, given their children a good education, and saved money. Ireland contains forty thousand farmers who occupy spots ranging from only one to two acres each. In France the subdivision of land into small parcels is equally extensive. Yet the occupants all live, and regard the being dispossessed of their small holdings as the misfortune of their lives. Why, then, cannot Americans who own two hundred or a thousand acres, make farming profitable? There is a controlling reason which has been repeatedly set forth—they plant too much, spreading their limited quantity of manure over too large a surface, thus impoverishing the land and wasting their labor. Eighty bushels of corn, and other grains in proportion, may be raised on one acre of land much easier than on two, and where land is so cultivated as to produce such crops it is constantly improving. When not so cultivated it is constantly depreciating.

Take, as an illustration, the case of two adjoining farms, that of A containing 150 acres, and that of B only 40. A has 40 acres of meadow, on which he cuts an annual average of 35 tons of hay, while B has only 15 acres of meadow, yielding him 2½ tons per acre, or 37½ in all. A plants 6 or 8 acres of corn every year, which yield him about 30 bushels to the acre, and his other field crops are in proportion, with proportionate results. On the other hand, B plants but 2 or 3 acres of corn, but he gathers from 75 to 80 bushels per acre, and is able to do all his work himself. A pays out $150 a year for help. He complains of hard times, of the scarcity of money, and talks of moving West, not because there is more money there, but because he can obtain more land. He seems unable to comprehend that he already has more than he is managing properly.

No such longings assail B. His little freehold is too precious in his eyes to be alienated—he is satisfied that no change could be for the better. His farm is constantly improving in value, while the other is annually decreasing. The latter is worked by a *skinner*, the former by a *farmer*. The skinner ploughs and plants indiscriminately. Around his barns the manure heaps lie unused from year to year. He reads no agricultural books, takes no agricultural paper, sells all his best stock, and is thus compelled to keep that which is so worthless as to be unsalable. But B suffers no particle of manure to remain unused, and ploughs only such land as he can thoroughly enrich. He sells his poorest stock and keeps only the best. He takes the best agricultural periodicals, and though he may read in them the most glowing accounts of distant lands at low prices, yet his affections are too firmly anchored in his little homestead for them to excite in him a single wish to leave it. It is not the *farming* that is unprofitable, but the *management.*

This recital revives once more the often mooted question of the comparative advantage of large or small farms. It cannot be denied that very large ones have been so managed by competent men as to yield enormous profits, while annually becoming more valuable. A single California farmer has harvested a crop of 400,000 bushels of potatoes, while a neighbor's crop amounted to 250,000 bushels. Another, at Los Angeles, with a vineyard of 35 acres, makes 35,000 gallons of wine annually. The wheat crop of that region reaches as high as 60 to 70 bushels per acre. The district around Sacramento is a vast wheat and barley field. Cattle require no housing as in the Atlantic States, but keep themselves the year round on the pastures.

In that State farming seems to be the sure road to fortune—the larger the well-managed farm the more rapid the accumulation. It will surprise readers in the East to hear of the immense fruit orchards in California, which so far surpass any thing in the Atlantic States as to make our attempts in this way seem exceedingly small. The following, from the *California Farmer*, is an accurate account of the product of the orchard of Mr. G. C. Briggs, of Marysville:—

"I send you," says Mr. Briggs to the editor, "the number of pounds of fruit as kept in my daily record. I have annexed the price. We have sold as follows:—

Fruits.	No. Pounds.	Price. Cents.	Amount.
Cherries............	3,680	60	$2,208
Appricots...........	58,400	20	11,680
Plums..............	22,120	30	6,636
Peaches.............	763,600	8	61,088
Nectarines...........	93,400	8	7,472
Apples..............	225,000	13	29,250
Pears...............	11,300	15	1,695
Quinces.............	4,720	20	945
Figs................	6,300	20	1,260
Grapes..............	34,500	8	2,760
	12,223,020		$124,993"

Estimating it in bushels, it amounts to more than twenty thousand bushels. California well may claim the palm for her horticultural products, and though all her fruits may not equal those of our more northern clime, they certainly surpass them in abundance and size.

In our Western States there are farmers cultivating immense tracts of land with evident profit. There are farms of 10,000 acres, yielding 90,000 bushels of corn alone, with thousands of bushels of wheat, besides fattening great droves of hogs and cattle. But these facts are too well known to make a recapitulation necessary. These must be profitable undertakings, or their owners would abandon them.

"Commerce and manufactures," says another authority, "must ever be secondary to the cultivation of the soil. The latter is not only the most important of all human industrial pursuits, but is the only real source of wealth. Commerce produces nothing—its office being merely the barter of commodities. Whether this barter takes place between one country and another, or between individuals of the same country, it is but an exchange of equivalents. It is therefore to be regarded as a mere medium for the distribution or circulation of wealth, and not as in any way contributing to its existence or production. Then, as to manufactures, there is no matter produced which did not previously exist, their office being only to convert material previously existing into forms of greater utility and convenience. At first sight, mining may appear to have a greater claim to the production of wealth, but it does not, in reality, produce anything which did not before exist. Every pound of iron, silver, gold, or coal, existed in the bowels of the earth long before it was taken from them. Increase of matter proceeds from agriculture alone. The surplus of this over cost of production constitutes the only increase of real wealth or capital. Yet however true this may be, it must be remembered that commerce by the exchange of commodities, and manufactures by giving to the productions of agriculture a more useful form, are greatly conducive to the aggrandizement of nations, and to the convenience and comfort of their population."

The reader is probably going somewhere. He may possibly have various offers of employment or occupation, and find it difficult to decide between them. There is one rule which, in all such cases, will generally be found a safe one to follow—go where he is most wanted. It may not be that a man should always go where most clamored after, or even where most liberally paid, though either of these things affords an indication of where he ought to locate. In the foregoing pages, a shrewd and persevering man cannot fail to discover where he is most wanted—as well as the ways and means, how and where, to plant his stakes. If at a loss to decide, such hesitation cannot spring from lack of opportunities, but from their abundance. It is here apparent that everywhere around him there are openings that lie waiting to be appropriated. No other country in the world presents a similar spectacle. Throughout Europe, land is at a premium, and labor at a discount. Here the contrary is the rule—labor commands the premium, and land is held at the discount. Though slowly acquiring some of this foreign passion for the ownership of land, yet centuries must elapse before the American people become generally imbued with it. Of those who now hunger after freeholds, too many are mere speculators. But recent legislation has placed beyond their grasp the largest portion of the public domain. It remains for the actual settler to crowd them altogether from the arena.

THE END.

BOOKS FOR FARMERS.

The subscriber has constantly on hand all **Standard Works** on

AGRICULTURE, HORTICULTURE, LANDSCAPE GARDENING, CATTLE AND THEIR DISEASES, THE HORSE, THE STABLE.

To which he would invite attention.

Persons living at a distance, can have any book sent to them by forwarding the advertised price.

JAMES MILLER, Publisher,
522 Broadway, N. Y.

BUY THE BEST!—THE PREMIUM

THRESHING MACHINE

The RAILWAY HORSE-POWER, which has repeatedly taken the **First Premium at the New York State Fair,** and has *never failed to do so over all its competitors* wherever exhibited by us in competition with others, running with low elevation and slow travel of team!

Combined Threshers and Cleaners, Threshers, Separators, Fanning Mills, Wood Saws, &c., all of the best in market. THE THRESHER AND CLEANER received the *first premium* at the Ohio State Fair, 1863, runs easy, separates the grain clean from the straw, cleans quite equal to the best of Fanning Mills, leaving the grain fit for mill or market.

☞ For Price and Description send for Circulars, and satisfy yourself before purchasing.

R. & M. HARDER,
Cobleskill, Schoharie Co., N. Y.

www.ingramcontent.com/pod-product-compliance
Lightning Source LLC
LaVergne TN
LVHW010159110826
845151LV00002B/553
9781425536206